Symmetric and Alternating Groups as Monodromy Groups of Riemann Surfaces I: Generic Covers and Covers with Many Branch Points

MEMOIRS
of the
American Mathematical Society

Number 886

Symmetric and Alternating Groups as Monodromy Groups of Riemann Surfaces I: Generic Covers and Covers with Many Branch Points

with an Appendix by R. Guralnick and R. Stafford

Robert M. Guralnick
John Shareshian

September 2007 • Volume 189 • Number 886 (third of 4 numbers) • ISSN 0065-9266

American Mathematical Society
Providence, Rhode Island

2000 *Mathematics Subject Classification.* Primary 14H30, 20B15.

Library of Congress Cataloging-in-Publication Data
Guralnick, Robert M., 1950–
　Symmetric and alternating groups as monodromy groups of Riemann surfaces I: generic covers and covers with many branch points / Robert M. Guralnick, John Shareshian; with an appendix by R. Guralnick and J. Stafford.
　　p. cm. — (Memoirs of the American Mathematical Society, ISSN 0065-9266 ; no. 886)
　"September 2007, volume 189, number 886 (third of 4 numbers)."
　Includes bibliographical references.
　ISBN 978-0-8218-3992-8 (alk. paper)
　1. Permutation groups.　2. Curves　3. Monodromy groups.　4. Riemann surfaces.　5. Symmetry groups.　I. Title.
QA175.G87　2007
512′.21—dc22
　　　　　　　　　　　　　　　　　　　　　　　　　　　　　　　　　　2007060802

Memoirs of the American Mathematical Society

This journal is devoted entirely to research in pure and applied mathematics.

Subscription information. The 2007 subscription begins with volume 185 and consists of six mailings, each containing one or more numbers. Subscription prices for 2007 are US$649 list, US$519 institutional member. A late charge of 10% of the subscription price will be imposed on orders received from nonmembers after January 1 of the subscription year. Subscribers outside the United States and India must pay a postage surcharge of US$38; subscribers in India must pay a postage surcharge of US$43. Expedited delivery to destinations in North America US$53; elsewhere US$130. Each number may be ordered separately; *please specify number* when ordering an individual number. For prices and titles of recently released numbers, see the New Publications sections of the *Notices of the American Mathematical Society*.

Back number information. For back issues see the *AMS Catalog of Publications*.

Subscriptions and orders should be addressed to the American Mathematical Society, P. O. Box 845904, Boston, MA 02284-5904, USA. *All orders must be accompanied by payment*. Other correspondence should be addressed to 201 Charles Street, Providence, RI 02904-2294, USA.

Copying and reprinting. Individual readers of this publication, and nonprofit libraries acting for them, are permitted to make fair use of the material, such as to copy a chapter for use in teaching or research. Permission is granted to quote brief passages from this publication in reviews, provided the customary acknowledgment of the source is given.

Republication, systematic copying, or multiple reproduction of any material in this publication is permitted only under license from the American Mathematical Society. Requests for such permission should be addressed to the Acquisitions Department, American Mathematical Society, 201 Charles Street, Providence, Rhode Island 02904-2294, USA. Requests can also be made by e-mail to reprint-permission@ams.org.

Memoirs of the American Mathematical Society is published bimonthly (each volume consisting usually of more than one number) by the American Mathematical Society at 201 Charles Street, Providence, RI 02904-2294, USA. Periodicals postage paid at Providence, RI. Postmaster: Send address changes to Memoirs, American Mathematical Society, 201 Charles Street, Providence, RI 02904-2294, USA.

　　　　　　　© 2007 by the American Mathematical Society. All rights reserved.
　　　　　　Copyright of this publication reverts to the public domain 28 years
　　　　　　　　after publication. Contact the AMS for copyright status.
　This publication is indexed in *Science Citation Index*®, *SciSearch*®, *Research Alert*®,
　　CompuMath Citation Index®, *Current Contents*®/*Physical, Chemical & Earth Sciences*.
　　　　　　　　　　Printed in the United States of America.

　　　　∞ The paper used in this book is acid-free and falls within the guidelines
　　　　　　　established to ensure permanence and durability.
　　　　　　　　Visit the AMS home page at http://www.ams.org/

　　　　　　　　　10 9 8 7 6 5 4 3 2 1　　12 11 10 09 08 07

Contents

Chapter 1.	Introduction and statement of main results	1
1.1.	Five or more branch points	8
1.2.	An n-cycle	9
1.3.	Asymptotic behavior of the genus for actions on k-sets	10
1.4.	Galois groups of trinomials	10
Chapter 2.	Notation and basic lemmas	13
Chapter 3.	Examples	21
Chapter 4.	Proving the main results on five or more branch points - Theorems 1.1.1 and 1.1.2	29
Chapter 5.	Actions on 2-sets - the proof of Theorem 4.0.30	33
Chapter 6.	Actions on 3-sets - the proof of Theorem 4.0.31	65
Chapter 7.	Nine or more branch points - the proof of Theorem 4.0.34	79
Chapter 8.	Actions on cosets of some 2-homogeneous and 3-homogeneous groups	81
Chapter 9.	Actions on 3-sets compared to actions on larger sets	89
Chapter 10.	A transposition and an n-cycle	97
Chapter 11.	Asymptotic behavior of $g_k(E)$	103
Chapter 12.	An n-cycle - the proof of Theorem 1.2.1	107
Chapter 13.	Galois groups of trinomials - the proofs of Propositions 1.4.1 and 1.4.2 and Theorem 1.4.3	117
Appendix A.	Finding small genus examples by computer search - by R. Guralnick and R. Stafford	123
A.1.	Description	123
A.2.	$n = 5$ and $n = 6$	123
A.3.	$5 \leq r \leq 8, 7 \leq n \leq 20$	124
A.4.	$r < 5$	125
Appendix.	Bibliography	127

Abstract

We consider indecomposable degree n covers of Riemann surfaces with monodromy group an alternating or symmetric group of degree d. We show that if the cover has five or more branch points then the genus grows rapidly with n unless either $d = n$ or the curves have genus zero, there are precisely five branch points and $n = d(d-1)/2$. Similarly, if there is a totally ramified point, then without restriction on the number of branch points the genus grows rapidly with n unless either $d = n$ or the curves have genus zero and $n = d(d-1)/2$. One consequence of these results is that if $f : X \to \mathbb{P}^1$ is indecomposable of degree n with X the generic Riemann surface of genus $g \geq 4$, then the monodromy group is S_n or A_n (and both can occur for n sufficiently large). We also show if that if $f(x)$ is an indecomposable rational function of degree n branched at 9 or more points, then its monodromy group is A_n or S_n. Finally, we answer a question of Elkies by showing that the curve parameterizing extensions of a number field given by an irreducible trinomial with Galois group H has large genus unless $H = A_n$ or S_n or n is very small.

Received by the editor September 20, 2006.

1991 *Mathematics Subject Classification*. 14H30,20B15.

Key words and phrases. Monodromy group, Riemann surface, genus, generic curve, symmetric group, alternating group.

R. Guralnick gratefully acknowledges the support of the NSF (grants DMS-9970305 and DMS-0140578).

J. Shareshian gratefully acknowledges the support of the NSF (grants DMS-0070757 and DMS-0030483).

CHAPTER 1

Introduction and statement of main results

Let X, Y be Riemann surfaces. Let $f : X \to Y$ be a nonconstant rational map of degree n. Suppose that X has genus g. Let G be the monodromy group of this cover (i.e. G is the Galois group of the Galois closure of $\mathbb{C}(X)/\mathbb{C}(Y)$). So G is a transitive group of degree n.

The main problem we are interested in is the following: if we fix the genus of X, what restrictions are placed on G? See [**GuTh**] and [**GuNe**] for more background on this problem.

It is easy to show (see the references cited above or [**Gu2**] for analogous results in all characteristics) that the critical case to study is when $Y = \mathbb{P}^1$ and f is indecomposable. In that case G is a primitive permutation group of degree n.

Let $B \subset \mathbb{P}^1$ the set of branch points of the cover above (i.e. the points where the cover is ramified). Set $r = |B|$.

Since the fundamental group of $\mathbb{P}^1 \setminus B$ is a group generated by r elements with the single relation that the product is 1, using Riemann's Existence Theorem, we see that the existence of a cover gives rise to elements $x_1, \ldots, x_r$ in G such that $G = \langle x_1, \ldots, x_r \rangle$, $x_1 \cdots x_r = 1$. Moreover, G is a primitive permutation group of degree n. Let Ω denote this G-set of size n. We have the Riemann-Hurwitz formula:

$$2(n + g - 1) = \sum_{i=1}^{r} \operatorname{ind}(x_i, \Omega),$$

where $\operatorname{ind}(x, \Omega) = n - \operatorname{orb}(x, \Omega)$ and $\operatorname{orb}(x, \Omega)$ is the number of orbits of x on Ω.

We shall study this problem from a group theoretic point of view. So we can translate the cover into "monodromy data" – i.e. a triple (G, H, E) where G is the monodromy group, H is a point stabilizer in the permutation representation and E is an r-tuple of elements of G which generate G and whose product is 1. We can then define the genus $g(G, H, E)$ by the formula above.

Indeed, Ree [**Re**] observed that $g(G, H, E) \geq 0$ by noting that it is the genus of a Riemann surface. Subsequent purely algebraic proofs were given in [**FeLySc**] and [**Sc**]. The latter generalization has proved to be very useful in many aspects of group theory.

In [**GuTh**], it was conjectured that for a fixed g, the noncyclic and nonalternating composition factors of G would all lie in some finite set $\mathcal{E}(g)$. This is commonly called the Guralnick-Thompson conjecture. This was reduced to the case that G is almost simple (i.e. G has a unique minimal normal subgroup that is a nonabelian simple group) in [**GuNe**]. After a series of papers by various authors ([**GuTh, As, Gu1, Sh1, Sh2, LiSa, Ne1, Ne2, LiWP, LiSh**]), this was finally settled in [**FrMa**].

However, one really wants much stronger information than restrictions on composition factors. Indeed we have the following conjecture of the first author:

CONJECTURE 1.0.1. *There exists a real valued function N such that if $f : X \to \mathbb{P}^1$ is an indecomposable cover of degree n with X of genus g, then one of the following holds for the monodromy group G:*

(1) $n \leq N(g)$;
(2) G is almost simple with socle A_d and $n = d$;
(3) $g \leq 1$ and one of:
   (a) $g = 0$, G is almost simple with socle A_d and $n = d(d-1)/2$;
   (b) $g \leq 1$, the socle of G is $A_d \times A_d$ and $n = d^2$;
   (c) $g \leq 1$, $n = p$ or p^2 for some prime p, G has a unique normal subgroup N of order n and G/N is cyclic of order $1, 2, 3, 4$ or 6.

The authors together with Frohardt and Magaard want to classify all such covers with X of genus at most 2 (see [**FGM**] for some results along these lines). Clearly for both of these problems, one needs to deal with symmetric groups (and in a certain sense this is one of the biggest obstacles to obtaining these results – the other being considering groups close to $S_d \wr H$ where H is cyclic of order 2 or 3 or S_3).

Since alternating groups are allowed as composition factors, the case of G a symmetric or alternating group was not studied very much in the papers mentioned above. In fact, the methods used are really not applicable. The main point in much of the work cited above was to use the fact that an element in an almost simple permutation group other than an alternating or symmetric group does not fix very many points (see [**LiSa**] and [**GuMa**]). This is not the case when G is alternating or symmetric – particularly when the permutation action is on the set of k-sets for k reasonably small. One sees that for example a transposition in S_d moves only $2(d-2)$ subsets of size 2 and fixes the rest – i.e. its fixed point ratio on the set of subsets of size 2 is $1 - 4(d-2)/d(d-1)$ or roughly $1 - 4/n^{1/2}$ where n is the number of 2-sets.

Thus, we need to use other methods to attack this problem. There are two key observations that make this problem feasible. Let $G \in \{A_d, S_d\}$ with $d \geq 5$ and let E be an (ordered) generating (multi)set for G with product 1. We set $g_k(E)$ to be the genus of (G, H_k, E) where H_k is the stabilizer of a k-set for $1 \leq k < d/2$. Then $g_k(E) \leq g_{k+1}(E)$ (see Lemma 2.0.12). Thus, we will focus our attention on g_2. Most of the work shows that in most cases g_2 is much bigger than g_1. One can then show that g_k grows rapidly for $k \geq 3$. Using some character theory, one can show that the other possibilities for H almost always yield large genus.

In this paper, we focus on the cases where either $r \geq 5$ or one of the generators is an n-cycle. These are two very important cases. We will deal with the cases $r = 3, 4$ in a sequel - see the appendix for results for $r = 3, 4$ with $d \leq 50$.

The case of an n-cycle is easier to deal with than the general case. It is quite important for it corresponds to covers where there is a totally ramified point. Moreover, it also occurs in the description of the curve parameterizing trinomials with a given Galois group. We describe the results below. We thank Noam Elkies for suggesting this problem to us.

Before going into a detailed description of the main results, we point out two applications of our results. The first depends on results in [**GuNe**] and [**GuMa**] and the results of this paper:

THEOREM 1.0.2. *Let $f : \mathbb{P}^1 \to \mathbb{P}^1$ be a degree n nonconstant indecomposable rational function with 9 or more branch points. Then the monodromy group is S_n or A_n.*

The next application is to covers of the Riemann sphere by the generic Riemann surface of genus g. This was a problem first studied by Zariski who proved that if $g > 6$, then a cover of the Riemann sphere by the generic Riemann surface of genus g cannot have solvable monodromy group. We make this concept more precise.

Suppose we are given monodromy data (G, H, E) of genus g. Let $\mathcal{M}_g$ denote the moduli space of genus g Riemann surfaces. Let $\mathcal{M}_g(G, H, E)$ be the set of genus g curves X such that there exists $f : X \to \mathbb{P}^1$ with monodromy data (G, H, E). It can be shown that this is a quasiprojective subvariety of $\mathcal{M}_g$. If this subvariety has full moduli dimension (i.e. $3g - 3$ for $g \geq 2$), we say that the generic curve of genus g admits a cover with that monodromy data.

It is well known that (S_n, S_{n-1}, E) with E a generating set consisting of the right number of transposition satisfies: $\mathcal{M}_g(G, H, E) = \mathcal{M}_g$ for $n > (g+1)/2$. In particular, every genus 6 curve can be realized with an S_4 cover of degree 4. Zariski proved that if $g > 6$ and $\mathcal{M}_g(G, H, E) = \mathcal{M}_g$, then G cannot be solvable. This was improved in [**GuNe**] to show that if we set $\mathcal{M}_g(G) = \bigcup_{(H,E)} \mathcal{M}_g(G, H, E)$, then $\cup_{G \text{ solvable}} \mathcal{M}_g(G)$ is contained in a proper subvariety for $\mathcal{M}_g$ for $g > 6$. Zariski also made the observations that one can reduce to the case that G is primitive on the cosets of H and that we only need consider covers to $\mathbb{P}^1$ (the generic curve of genus at least 2 has no nontrivial maps to a positive genus curve). He made the observation that in this case, $\dim \mathcal{M}_g(G, H, E) \leq |E| - 3$. See [**Fri2**] for more on this.

Combining our results with those in [**GuNe**] and [**GuMa**] yields:

THEOREM 1.0.3. *If $f : X \to \mathbb{P}^1$ is an indecomposable cover of degree n by the generic Riemann surface of genus $g > 3$, then $G = A_n, n \geq 2g + 1$ or S_n with $n > (g+1)/2$.*

Note that this result includes Zariski's result on solvable monodromy groups. If $g = 3$, there are a few more possibilities (see [**GuMa**]). If $g = 2$, there are considerably more possibilities. Zariski conjectured that as long as $|E| \geq 3g$, the generic curve is realized. This in fact is false (cf. [**FriGu**]) – but as far as we know only false for $g = 2$ in two cases.

Magaard and Völklein [**MaVo**] have show that in fact (A_n, A_{n-1}, E) has full moduli dimension for $n \geq 2g + 1$ and E consisting of the appropriate number of 3-cycles (or involutions moving four points). Also Artebani and Pirola have shown that for $n \geq 12g + 4$, every curve of genus g admits a degree n cover to the Riemann sphere in which every branch point is triple point [**ArPi**].

In the remainder of this chapter, we give more technical statements of our main results. Before doing that, we give an outline of the rest of the paper.

In Chapter 2 we introduce the notation that will be used in all subsequent parts of the paper and prove some basic results. Some of these are simple counting or arithmetic results which will be used repeatedly, sometimes without reference. Others express some of the key ideas of our approach. In particular, Lemma 2.0.11 says that $2g(G, H, E)$ is the dimension of the trivial subspace of the $\mathbb{C}[H]$-module determined by the action of H on the first homology group of a cover of $\mathbf{P}^1$ with monodromy data $(\rho(G), 1, \rho(E))$, where $\rho : G \to S_{|G|}$ is a regular representation.

This has several consequences. One is the above claim that $g_k(E) \leq g_{k+1}(E)$ whenever $k < \frac{n}{2}$ (Lemma 2.0.12). This follows from well known facts about the representation theory of S_n. Moreover, using Frobenius reciprocity, in Lemma 2.0.13 we obtain lower bounds on $g(G, H, E)$ in terms of the number of orbits of H on k-sets on $g_k(E)$ for small k. Later, this will allow us to reduce the examination of $g(G, H, E)$ for arbitrary H to the case where H is 3-homogeneous. Less central but key to our final results are Proposition 2.0.16 and Corollaries 2.0.17 and 2.0.18, which allow us to show that for certain H, E for which calculations give $g(G, H, E)$ small, there cannot exist E of the types under consideration with $A_n \leq \langle E \rangle$.

In Chapter 3, we describe several infinite classes of r-tuples $E = (x_1, \ldots, x_r)$ from S_n such that $\prod_{i=1}^r x_i = 1$, $A_n \leq \langle E \rangle$ and $g_2(E) = 0$. In all cases, we have $r \leq 5$. The presence of these examples gives some evidence that the tedious cases analysis that appears later in this paper is likely unavoidable. With the exception of some conjectured additional examples at the end of the chapter, all of these examples are derived by making small adjustments to two types of triples. The first of these types is a triple $E = (t, y, z)$, where t is a transposition and z is an n-cycle. Thus y has two cycles of lengths $a, n - a$, and it is straightforward to show that $\langle E \rangle = S_n$ if and only if $\gcd(a, n) = 1$. These triples are examined in detail in Chapter 10, where it is shown that they satisfy $g_2(E) = 0$. Lemmas 3.0.22 and 3.0.23 show that one can obtain from these systems further systems of sizes four and five satisfying $g_2(E) = 0$ by replacing either or both of y, z by an appropriate pair of involutions. These systems are described in Proposition 3.0.24. The second of these types is a triple F containing two involutions whose product has one or two cycles, along with the inverse of that product. Certainly such a triple generates only a dihedral group, but by finding an appropriate involution t and multiplying one element of F on the right by t and another on the left by t, one finds a triple of the desired type which generates A_n or S_n. Such examples are described in Propositions 3.0.25 and 3.0.28. Further classes that are obtained from those in Proposition 3.0.25 using Lemmas 3.0.22 and 3.0.23, are described in Propositions 3.0.26 and 3.0.27. Let us pause here to make the following conjecture (the analog of our main theorem when $|E| \geq 5$).

CONJECTURE 1.0.4. *There exists a constant $c > 0$ such that if $G \in \{A_n, S_n\}$ with $n > 30$ and H is a maximal subgroup of G not containing A_{n-1} and (G, H, E) are monodromy data with $|E| \leq 4$ and $g(G, H, E) \leq \max\{2, cn\}$, then E is listed in Chapter 3 and H is the stabilizer of a 2-set.*

In Chapter 4 we prove our main theorems about coverings with five or more branch points. The first of these is Theorem 1.1.1, which says that if $G \in \{A_n, S_n\}$ and $E = (x_1, \ldots, x_r)$ is an r-tuple from $G \setminus \{1\}$ with $G = \langle E \rangle$, $\prod_{i=1}^r x_i = 1$ and $r \geq 5$, then one of the conditions

- $g_k(E) \geq \max\{cn, 2, \frac{r}{3}\}$ for some constant c which does not depend on G, k or E,
- $G = S_n$, $g_k(E) = 0$, $r = 5$, $k = 2$ and E is one of the 5-tuples described in Proposition 3.0.24, or
- E is one of finitely many r-tuples described in the appendix

must hold. (Note that the condition $g_k(E) > \frac{r}{3}$ implies that if H is the stabilizer of a k-set in G, then $\mathcal{M}_{g_k(E)}(G, H, E)$ is not dense in $\mathcal{M}_{g_k(E)}$.) The second main theorem is Theorem 1.1.2, which says that if $G \in \{A_n, S_n\}$, H is a maximal subgroup of G (with $(G, H) \neq (S_n, A_n)$) and $E = (x_1, \ldots, x_r)$ is an r-tuple from G satisfying the conditions described for Theorem 1.1.1, then one of the conditions

- H is a point stabilizer in the natural action of G,
- $g(G, H, E) > \max\{cn, 2\}$ for some constant c which does not depend on G, H or r,
- $G = S_n$, H is the stabilizer of a 2-set in the natural action of G, $r = 5$ and (G, H, E) is one of the 5-tuples described in Proposition 3.0.24, or
- (G, H, E) is one of finitely many triples described in the appendix

must hold.

The hard work in our proof is delayed until Chapters 5 and 6. In Chapter 5, we prove Theorem 4.0.30, which says that if G and E are as in Theorem 1.1.1 and $n > 20$ then either

- $g_2(E) - g_1(E) > \max\{cn, 2\}$ for some constant c which is independent of G and E and $g_2(E) > \frac{r}{3}$, or
- $G = S_n$, $r = 5$, $g_2(E) = 0$ and E is one of the 5-tuples described in Proposition 3.0.24.

The proof of Theorem 4.0.30, which we now describe, is tedious. For $x \in S_n$ and $k \in [n]$, let $o_k(x)$ be the number of orbits of $\langle x \rangle$ on the set of k-subsets of $[n]$ and set

$$i_k(x) := \binom{n}{k} - o_k(x).$$

The Riemann-Hurwitz formula says that

$$2\left(g_k(E) + \binom{n}{k} - 1\right) = \sum_{x \in E} i_k(x).$$

After setting

$$\varepsilon_2(x) := \frac{i_2(x)}{\binom{n}{2}} - \frac{i_1(x)}{n},$$

direct calculation shows that

$$g_2(E) - g_1(E) = \frac{n-3}{2}(g_1(E) - 1) + \frac{n(n-1)}{4}\sum_{x \in E} \varepsilon_2(x).$$

We obtain lower bounds for $\varepsilon_2(x)$ which depend only on the number of short cycles in x in Lemma 5.0.40. Although straightforward to prove using Burnside's Lemma, this lemma is one of the key ingredients of our work. With it in hand, the proof of Theorem 4.0.30 is reduced to the careful but mathematically trivial examination of a large number of cases, depending on the number of transpositions in E and the number of elements of E having few fixed points (the elements just described being those for which $\varepsilon_2(x)$ is small).

In Chapter 6, we prove Theorem 4.0.31, which says that if G and E are as in Theorem 1.1.1 and $n > 20$ then

- $g_3(E) - g_2(E) \geq \max\{\frac{n-10}{3}, cn^2\}$

for some constant c which does not depend on G or E (so $g_3(E) - g_2(E) > 2$). This result is proved in a similar manner to that which was used to prove Theorem

4.0.30, although here the more tedious work is in getting lower bounds on

$$\varepsilon_{2,3}(x) := \frac{i_3(x)}{\binom{n}{3}} - \frac{i_2(x)}{\binom{n}{2}}$$

(see Lemmas 6.0.52 and 6.0.53) rather than examining various conditions on E.

Let us now return to our proof of Theorems 1.1.1 and 1.1.2 in Chapter 4. Theorem 1.1.1 under the assumption $n > 20$ follows quickly from Theorems 4.0.30 and 4.0.31. Indeed, by Riemann's Existence Theorem, we have $g_1(E) \geq 0$ for all E under consideration, so $g_2(E) > \max\{cn, 2\frac{r}{3}\}$ for all E not described in Proposition 3.0.24 by Theorem 4.0.30. For those E described in the proposition, we have $g_3(E) > \max\{\frac{n-10}{3}, c'n^2\}$. Theorem 1.1.1 for $n > 20$ now follows from Lemma 2.0.12, and the case $n < 20$ is handled by direct inspection in the appendix.

To prove Theorem 1.1.2 under the assumption that $n > 20$, we note first that by Theorem 1.1.1, we may assume that H is transitive. We then apply the above-mentioned Lemma 2.0.13, which we describe in more detail here. For $H \leq S_n$ and $k \in [n]$, let $o_k(H)$ be the number of orbits of H on k-subsets of $[n]$. Lemma 2.0.13 says that

$$(1.1) \qquad g(G, H, E) \geq \sum_{k=1}^{\lfloor \frac{n}{2} \rfloor} (o_k(H) - o_{k-1}(H))(g_k(E) - g_{k-1}(E)).$$

The quantity $o_k(H)$ can be computed directly when k is small and H is a maximal imprimitive subgroup of G, and such computations (along with (1.1) and Theorems 4.0.30 and 4.0.31) show that the theorem holds when H is such a group. We are left with the case that H is primitive. Using (1.1) and Theorem 4.0.31, we see that the claim of the Theorem holds when $o_3(H) - o_2(H) > 0$. A result of P. Cameron, P. Neumann and J. Saxl says that if $o_3(H) = o_2(H)$ then H is 3-homogeneous. All 3-homogeneous primitive groups are known. In Chapter 8, with the list of such groups in hand, we prove Theorem 4.0.33. This theorem says, with no restrictions on n and only finitely many explicitly listed exceptions (none of which allow $n > 20$), the claim of the theorem holds when H is 3-homogeneous.

It then remains to prove Theorems 1.1.1 and 1.1.2 when $n \leq 20$. The key point here is that we can show that the theorems essentially hold if $r \geq 9$. In Chapter 7 we prove Theorem 4.0.34, which says that if $G \in \{A_n, S_n\}$ with $n \geq 5$ and $E = (x_1, \ldots, x_r)$ is an r-tuple from $S_n \setminus \{1\}$ such that $\langle E \rangle = G$, $\prod_{i=1}^r x_i = 1$ and $r \geq 9$, then

- $g_2(E) - g_1(E) > 2$ and $g_2(E) > \frac{r}{3}$.

With this result in hand, we can proceed as we did in the case $n > 20$ to show that the claims of the theorems hold if H is not primitive and 2-homogeneous. All primitive groups of degree at most 20 are known, and in Chapter 8 we prove Proposition 4.0.35, which gives the (short) list of all triples (G, H, E) such that $G \in \{A_n, S_n\}$ with $n \leq 20$, $H \neq A_n$ is a 2-homogeneous maximal subgroup of G, $|E| \geq 5$ and $g(G, H, E) \leq \max\{2, \frac{|E|}{3}\}$. In none of the listed cases do we have $|E| \geq 9$ and $n \geq 7$. Moreover, in the only cases in which we can have $|E| \geq 9$ and $n \geq 5$, we must have $(G, H) \in \{(A_6, L_2(5)), (S_6, PGL_2(5))\}$ and $|E| \leq 14$. There are now only finitely many cases remaining to be examined, and these cases are handled by computer calculation, with results given in the appendix. Thus our main results are proved.

In Chapter 9, we prove Lemma 9.0.64, which gives a lower bound on
$$\frac{i_3(x)}{\binom{n}{3}} - \frac{i_k(x)}{\binom{n}{k}}$$
for $x \in S_n$ and $4 \leq k \leq n-4$. This bound used in Chapter 10 to prove Theorem 10.0.66, which says that if $E = (t, z, y)$, where $t \in S_n$ is a transposition, z is an n-cycle with $\langle t, z \rangle = S_n$ and $y = (tz)^{-1}$ then
- $g_1(E) = g_2(E) = 0$,
- there is some constant $c > 0$ such that if $n > 7$ and $3 \leq k \leq n-3$ then
$$\frac{g_k(E)}{\binom{n}{k}} > \frac{c}{n},$$
- if $n > 6$ then, with finitely many explicitly listed exceptions, $g_3(E) > 2$, and
- if $n > 9$ then $g_4(E) - g_3(E) > 2$.

This result is key for those of the two following chapters. In Chapter 11, we examine the asymptotic behavior of $g_k(E)$. We prove Theorem 1.3.1, which says that for each $k > 1$ there exist constants $c_{k,1}, c_{k,2} > 0$ such that
- if E is one of our usual r-tuples with $r \geq 5$, then either $g_2(E) = 0$ and E is as described in Proposition 3.0.24 or
$$\frac{g_k(E)}{\binom{n}{k}} > \frac{c_{k,1}}{n},$$
and
- for each large enough n there exists a 5-tuple $E_n(k)$ such that
$$\frac{g_k(E_n(k))}{\binom{n}{k}} < \frac{c_{k,2}}{n}.$$

In Chapter 12, we prove another version of Theorem 1.1.2, removing the restriction $r \geq 5$ but assuming that E contains and n-cycle. This answers a question asked by P. Müller, who is interested in the classifying indecomposable rational functions that have the same Galois closure as a polynomial. The proof of the main result, Theorem 1.2.1, follows the same track as the proof of Theorem 1.1.2 - we prove versions of Theorems 4.0.30 and 4.0.31 for large enough n (we assume take $n > 20$ if $r > 3$ and $n > 47$ if $r = 3$), then reduce to the case where H is 3-homogeneous and apply our knowledge of 3-homogeneous groups. This leaves us with finitely many cases, which are handled by computer, as described in the appendix. Note that the case $r \geq 5$ has already been addressed in Theorem 1.1.2. Moreover, using Lemma 3.0.22, the case $r = 4$ is easily reduced to one which is addressed by Theorem 1.1.2. Thus our work in this chapter is concentrated on the case $r = 3$. We encounter here the various classes of E for which $g_2(E) = 0$ which are described in Chapter 3.

In Chapter 13, we apply the methods used to prove our main results to a problem in inverse Galois theory posed by N. Elkies. Let K be a number field. Let $z \in S_n$ be the n-cycle $(1, 2, \ldots, n)$. For $a \in [n]$, let t_a be the transposition $(1, a+1)$. Let $y_a = (t_a z)^{-1}$. As shown in Chapter 10, we have $\langle t_a, z, y_a \rangle = S_n$ if and only if $\gcd(a, n) = 1$. Assume that $\gcd(a, n) = 1$. For $H \leq S_n$ let $\mathcal{F}(H)$ be the set of trinomials $x^n + \alpha x^a + \beta$ over K whose Galois group is contained in H. Define an equivalence relation on $\mathcal{F}(H)$ by setting f equivalent to g if there is some complex number τ such that the (complex) roots of f are $r_1, \ldots, r_n$ and the roots of g are $\tau r_1, \ldots \tau r_n$. As noted by Elkies (see [**El**]), there is a bijection between the equivalence classes in $\mathcal{F}(H)$ and the set of K-rational points on a Riemann surface $X = X(H, a)$. Also, there is an analytic map $f : X \to \mathbf{P}^1$ with monodromy data $(S_n, H, \{t_a, z, y_a\})$.

By Faltings' Theorem (that is, Mordell's conjecture, see [**Fa**]), if $g(X) \geq 2$ then there are only finitely many K-rational points on X and therefore only finitely many classes of extensions of the type described above. Our main results, Propositions 1.4.1 and 1.4.2 and Theorem 1.4.3, give a complete list of those H and a for which $g(X) \leq 2$. This lists consists of all (H, a) such that H is one of S_n, A_n, S_{n-1}, $S_{n-2} \times S_2$ or S_{n-2}, along with finitely many other pairs.

Acknowledgments: As mentioned above, there is an appendix to this paper, written by the first named author and Richard Stafford of the National Security Agency. It contains a description of the methods used to obtain by computer calculations all triples of monodromy data (G, H, E) such that $g(G, H, E)$ is small, under various upper bounds on n and $|E|$, along with the results of these calculations. Both authors of the main body of the paper thank Dr. Stafford for his work on this integral part of the project. We also thank William Beckner for some very helpful comments and Noam Elkies for suggesting the problem on Galois groups of trinomials that is discussed in Chapter 13.

1.1. Five or more branch points

THEOREM 1.1.1. *Let $G \in \{A_n, S_n\}$ with $n \geq 5$. Assume $2 \leq k \leq n-2$, and let H be the stabilizer of a k-set in the natural action of G on $[n]$. Let $E = (x_1, \ldots, x_r)$ be an r-tuple of nonidentity elements of G such that $G = \langle E \rangle$ and $\prod_{i=1}^r x_i = 1$. Assume that $r \geq 5$. Then there is some constant $c > 0$, which does not depend on G, H or n, such that one of the following conditions holds.*

(1) *We have*
$$g(G, H, E) > \max\{cn, 2, \frac{r}{3}\}.$$
(2) *$G = S_n$, $k = 2$, $g(G, H, E) = 0$ and one of the following two conditions holds.*
   (a) *The five elements of E have cycle shapes $1^{n-2}2^1$, $1^3 2^{\frac{n-3}{2}}$, $1^1 2^{\frac{n-1}{2}}$, $1^1 2^{\frac{n-1}{2}}$, $1^1 2^{\frac{n-1}{2}}$.*
   (b) *The five elements of E have cycle shapes $1^{n-2}2^1$, $1^2 2^{\frac{n-2}{2}}$, $1^2 2^{\frac{n-2}{2}}$, $1^2 2^{\frac{n-2}{2}}$, $2^{\frac{n}{2}}$.*
(3) *We have $r = 5$ and $n \in \{5, 6\}$.*
(4) *We have $r > 5$, $n \in \{5, 6\}$, and (G, H, E) is one of the finitely many triples described in Theorem A.2.1.*
(5) *We have $7 \leq n \leq 20$ and (G, H, E) is one of the finitely many triples described in Theorems A.3.1 and A.3.2.*

Moreover, for each case that is described in Conditions (2) and (4), there exist r-tuples $E = (x_1, \ldots, x_r)$ of nonidentity elements of G with $\langle E \rangle = G$ and $\prod_{i=1}^r x_i = 1$ which are described by the given case.

THEOREM 1.1.2. *Let $G \in \{A_n, S_n\}$ with $n \geq 5$. Let $E = (x_1, \ldots, x_r)$ be an r-tuple of nonidentity elements of G such that $G = \langle E \rangle$ and $\prod_{i=1}^r x_i = 1$. Let $H \neq A_n$*

be a maximal subgroup of G. Assume $r \geq 5$. Then there is some constant $c > 0$, which does not depend on G, H or n, such that one of the following conditions holds.

(1) H is the stabilizer of a point in the natural action of G.
(2) We have
$$g(G, H, E) > \max\{cn, 2\}.$$
(3) $G = S_n$, H is the stabilizer of a 2-set in the natural action of G, $g(G, H, E) = 0$ and condition (2) of Theorem 1.1.1 holds.
(4) One of the cases listed in Conditions (2),(4) and (5) of Theorem 1.1.1 holds.
(5) We have $n \in \{5, 6\}$.

1.2. An n-cycle

THEOREM 1.2.1. Let $G \in \{A_n, S_n\}$ with $n \geq 5$. Let $E = (x_1, \ldots, x_r)$ be an r-tuple of nonidentity elements of G such that $G = \langle E \rangle$ and $\prod_{i=1}^r x_i = 1$. Let $H \neq A_n$ be a maximal subgroup of G. Assume that x_r is an n-cycle. Then there is some constant $c > 0$, which does not depend on G, H or n such that one of the following conditions holds.

(1) H is the stabilizer of a point in the natural action of G.
(2) We have $g(G, H, E) > \max\{2, cn\}$.
(3) H is the stabilizer of a 2-set in the natural action of G, $g(G, H, E) = 0$ and one of the following conditions holds.
   (a) $E \setminus \{x_r\}$ consists of one element of shape $1^{n-2}2^1$, one element of shape $1^3 2^{\frac{n-3}{2}}$ and one element of shape $1^1 2^{\frac{n-1}{2}}$.
   (b) $E \setminus \{x_r\}$ consists of one element of shape $1^{n-2}2^1$ and two elements of shape $1^2 2^{\frac{n-2}{2}}$.
   (c) $E \setminus \{x_r\}$ consists of one element of shape $1^{n-2}2^1$ and one element of shape $a^1 b^1$ with $\gcd(a, b) = 1$.
   (d) $E \setminus \{x_r\}$ consists of one element of shape $1^3 2^{\frac{n-3}{2}}$ and one element of shape $1^1 2^{\frac{n-5}{2}} 4^1$.
   (e) $E \setminus \{x_r\}$ consists of one element of shape $1^3 2^{\frac{n-3}{2}}$ and one element of shape $2^{\frac{n-3}{2}} 3^1$.
   (f) $E \setminus \{x_r\}$ consists of one element of shape $1^2 2^{\frac{n-2}{2}}$ and one element of shape $1^2 2^{\frac{n-6}{2}} 4^1$.
   (g) $E \setminus \{x_r\}$ consists of one element of shape $1^2 2^{\frac{n-2}{2}}$ and one element of shape $1^1 2^{\frac{n-4}{2}} 3^1$.
   (h) $E \setminus \{x_r\}$ consists of one element of shape $1^1 2^{\frac{n-1}{2}}$ and one element of shape $1^3 2^{\frac{n-7}{2}} 4^1$.
   (i) $E \setminus \{x_r\}$ consists of one element of shape $1^1 2^{\frac{n-1}{2}}$ and one element of shape $1^2 2^{\frac{n-5}{2}} 3^1$.
(4) We have $n \leq 20$, $r \leq 4$ and (G, H, E) is one of the finitely many triples described in Theorem A.4.1.
(5) We have $r = 3$ and $n \leq 9$.

Moreover, for each case that is described in Condition (3), there exist r-tuples $E = (x_1, \ldots, x_r)$ of nonidentity elements of G with $\langle E \rangle = G$ and $\prod_{i=1}^r x_i = 1$ which are described by the given case.

1.3. Asymptotic behavior of the genus for actions on k-sets

THEOREM 1.3.1. *Let $2 \leq k \in \mathbb{N}$. For $n \geq k$ and $G \in \{A_n, S_n\}$, let $G(k)$ be the stabilizer of a k-set in the natural action of G.*

(1) *There exists constant $c_{k,1}$ (which depends only on k) such that for all large enough n and all r-tuples $E = (x_1, \ldots, x_r)$ of nonidentity elements of S_n such that $r \geq 5$, $\langle E \rangle = G$ and $\prod_{i=1}^{r} x_i = 1$, either condition (2) of Theorem 1.1.2 holds or*

$$g(G, G(k), E) > \frac{c_{k,1}}{n} \binom{n}{k}.$$

(2) *There exists a constant $c_{k,2} > 0$ (which depends only on k) such that for all large enough n there exists a 5-tuple $E_n(k) = (x_1, \ldots, x_5)$ of nonidentity elements of S_n such that $\langle E \rangle = S_n$, $x_1 x_2 x_3 x_4 x_5 = 1$ and*

$$g(S_n, S_n(k), E_n(k)) < \frac{c_{k,2}}{n} \binom{n}{k}$$

1.4. Galois groups of trinomials

Define

$$\mathcal{S}(n, a) := \{H \leq S_n : g(X(H, a)) \leq 2\}.$$

For $i \in [n]$, let H_i be the stabilizer of the point i in the natural action of S_n. For distinct elements $i, j \in [n]$, let $H_{i,j} = H_i \cap H_j$ and let $H_{\{i,j\}}$ be the stabilizer of the set $\{i, j\}$ in the natural action of S_n. Define

$$\begin{aligned}\mathcal{S}'(n) &:= \{S_n, A_n\} \bigcup \{H_i : i \in [n]\} \bigcup \left\{H_{i,j} : (i,j) \in [n]^2, i \neq j\right\} \\ &\quad \bigcup \left\{H_{\{i,j\}} : \{i,j\} \in \binom{[n]}{2}\right\}.\end{aligned}$$

In Chapter 13 we will prove the following results.

PROPOSITION 1.4.1. *If $H \in \mathcal{S}'(n)$ then $g(X(H, a)) = 0$ for all a.*

PROPOSITION 1.4.2. *If $n \leq 4$ then $g(X(H, a)) = 0$ for all $H \leq S_n$ and all a.*

THEOREM 1.4.3. *Assume that $n \geq 5$ and $H \in \mathcal{S}(n, a) \setminus \mathcal{S}'(n)$.*

(1) If H is transitive then the triple (n, a, H) is listed in the table below.

n	a	H	Action of H	$g(X)$
5	4, 1	$Z_5.Z_4$	affine	0
5	3, 2	$Z_5.Z_4$	affine	1
5	4, 1	D_{10}	5-gon	0
5	3, 2	D_{10}	5-gon	1
5	4, 1	Z_5	cyclic	0
5	3, 2	Z_5	cyclic	1
6	5, 1	$(S_3 \times S_3).S_2 \cong E_9.D_8$	imprimitive	0
6	5, 1	$E_8.S_3$	imprimitive	0
6	5, 1	$E_9.Z_4$	$H < E_9.D_8$	2
6	5, 1	$E_9.D_4$	$H < E_9.D_8$	1
6	5, 1	$E_9.Z_2$	$H < E_9.D_8$	2
6	5, 1	$E_8.Z_3$	$H < E_8.S_3$	0
6	5, 1	S_4	cosets of D_4	2
6	5, 1	S_4	cosets of Z_4	1
6	5, 1	S_5	Sylow 5-subgroups	0
6	5, 1	A_5	Sylow 5-subgroups	0
7	6, 1	$GL_3(2) \cong L_2(7)$	1-spaces from $\mathbf{F}_2^3$	2
8	7, 1	$(S_4 \times S_4).Z_2$	imprimitive	0
8	5, 3	$(S_4 \times S_4).Z_2$	imprimitive	1
8	7, 1	$AGL_3(2)$	affine	2

(2) *If H is not transitive then the triple (n, a, H) is listed in the table below.*

n	a	H	Action of H	$g(X)$
5	all	A_4	fixes point	1
5	$4, 1$	$(S_3 \times S_2) \cap A_5$	fixes 2-set	1
5	$3, 2$	$(S_3 \times S_2) \cap A_5$	fixes 2-set	2
5	all	D_8	4-gon	0
5	$4, 1$	Z_4	$\langle (1234) \rangle$	1
5	$4, 1$	D_4	$\langle (12)(34), (14)(23) \rangle$	1
5	all	D_4	$\langle (12), (34) \rangle$	0
5	all	Z_6	$\langle (123)(45) \rangle$	1
5	$4, 1$	Z_3	$\langle (123) \rangle$	2
5	$4, 1$	Z_2	$\langle (12) \rangle$	1
5	$4, 1$	Z_2	$\langle (12)(34) \rangle$	2
6	$5, 1$	$S_3 \times S_3$	fixes 3-set	0
6	$5, 1$	$S_2 \times S_2 \times S_2$	$\langle (12), (34), (56) \rangle$	2
6	$5, 1$	A_5	fixes point	1
6	$5, 1$	$A_4 \times Z_2$	4-set $\times$ 2-set	2
6	$5, 1$	$D_8 \times Z_2$	4-gon $\times$ 2-set	1
6	$5, 1$	$S_3 \times A_3$	fixes 3-set	1
7	$6, 1$	$S_4 \times S_2$	fixes 4-set, 2-set	2
7	$6, 1$	$(S_3 \times S_3).S_2$	fixes point, $(3, 3)$ partition	2
7	all	A_6	fixes point	1
7	$6, 1$	$S_4 \times S_3$	fixes 3-set	0
7	$5, 4, 3, 2$	$S_4 \times S_3$	fixes 3-set	1
7	$6, 1$	$S_4 \times A_3$	fixes 4-set	2
8	$7, 5, 3, 1$	A_7	fixes point	2
8	$7, 1$	$S_5 \times S_3$	fixes 5-set	1
8	$5, 3$	$S_5 \times S_3$	fixes 5-set	2
8	$7, 1$	$S_4 \times S_4$	fixes 4-set	1
9	$8, 1$	$S_6 \times S_3$	fixes 3-set	1
9	$7, 2$	$S_6 \times S_3$	fixes 3-set	2
10	$9, 1$	$S_7 \times S_3$	fixes 3-set	2

CHAPTER 2

Notation and basic lemmas

In this chapter we introduce notation which we will use throughout the paper and prove some fundamental facts.

DEFINITION 2.0.4. Let n, k be positive integers with $k \leq n$. Let $x \in S_n$. Then
(1) $f_k(x)$ is the number of k-subsets of n fixed (setwise) by x,
(2) $o_k(x)$ is the number of orbits of $\langle x \rangle$ on $\binom{[n]}{k}$, and
(3) $i_k(x) := \binom{n}{k} - o_k(x)$.

Also, for a multiset E of nonidentity elements of S_n, $g_k(E)$ is the number satisfying

$$(2.1) \qquad 2\left[\binom{n}{k} + g_k(E) - 1\right] = \sum_{x \in E} i_k(x).$$

So, if $G \in \{S_n, A_n\}$, H is the stabilizer of a k-set in G and $E = (x_1, \ldots, x_r)$ is an r-tuple of nonidentity elements of G such that $\langle E \rangle = G$ and $\prod_{i=1}^r x_i = 1$, then $g_k(E)$ is the genus of any Riemann surface X for which there is an analytic map $f : X \to \mathbf{P}^1$ with monodromy data (G, H, E). It is clear from the definitions that

$$g_k(E) = g_{n-k}(E)$$

for all $k \in [n]$, and we will use this fact without explicit reference.

As our main method for showing that $g_k(E)$ is large is to compare $g_k(E)$ with some $g_j(E)$ which we already understand, the quantity defined below is very important.

DEFINITION 2.0.5. Let j, k, n be positive integers with $j, k \leq n$. Let $x \in S_n$. Then

$$\varepsilon_{j,k}(x) := \frac{i_k(x)}{\binom{n}{k}} - \frac{i_j(x)}{\binom{n}{j}}.$$

When $j = 1$ we will suppress j from the notation, so

$$\varepsilon_k(x) := \frac{i_k(x)}{\binom{n}{k}} - \frac{i_1(x)}{n}.$$

Direct calculation gives the following lemma, which will be used repeatedly.

LEMMA 2.0.6. *For any r-tuple E of nonidentity elements of S_n and any $j, k \in [n]$ we have*

$$(2.2) \qquad \frac{2(g_k(E) - 1)}{\binom{n}{k}} = \frac{2(g_j(E) - 1)}{\binom{n}{j}} + \sum_{x \in E} \varepsilon_{j,k}(x).$$

Equivalently, we have

$$\text{(2.3)} \quad \frac{g_k(E)}{\binom{n}{k}} = \frac{g_j(E)}{\binom{n}{j}} + \frac{1}{\binom{n}{k}} - \frac{1}{\binom{n}{j}} + \frac{1}{2}\sum_{x\in E}\varepsilon_{j,k}(x).$$

In particular, for each $k \in [n]$ we have

$$\text{(2.4)} \quad \frac{2(g_k(E)-1)}{\binom{n}{k}} = \frac{2(g_1(E)-1)}{n} + \sum_{x\in E}\varepsilon_k(x).$$

We will use the numbers $\varepsilon_{j,k}(x)$ and Lemma 2.0.6 in several ways. For any E which we must consider, Riemann's existence theorem shows that $g_1(E) \geq 0$. As we will see in Lemma 5.0.40, there is some constant C such that if $x \in S_n$ has enough fixed points then $\varepsilon_2(x) > \frac{C}{n}$. We can take C to be satisfactorily large, in the sense described below, if x does not have too many fixed points. So, if enough $x \in E$ have enough (but not too many) fixed points then the term $\sum_{x\in E}\varepsilon_2(x)$ in equation (2.4) will overwhelm the term $\frac{2(g_1(E)-1)}{n}$, which is at least $-\frac{2}{n}$, thus giving $g_2(E) > dn$ for some fixed constant $d > 0$ (which we will not bother to specify, since we make no effort to find the largest possible d). On the other hand, if some $x \in E$ has very few fixed points then $o_1(x)$ is reasonably small, so $i_1(x)$ is reasonably large. If $\varepsilon_2(x)$ is not too negative then $i_2(x)$ will also be reasonably large. It will follow that if E contains enough elements with few fixed points then $g_2(E) > dn$.

With these results in hand, lower bounds on $\varepsilon_{2,k}(x)$ for $k > 2$, as given in Lemma 6.0.54, along with equation (2.3) can be used to show the existence of some constant c such that

$$\frac{g_k(E)}{\binom{n}{k}} > \frac{c}{n}$$

for all k when n is large enough and $g_2(E) \neq 0$. In the (known) cases where n is large and $g_2(E) = 0$, we can show by direct calculation that

$$\frac{g_3(E)}{\binom{n}{3}} > \frac{c'}{n}$$

for some constant $c' > 0$. Using equation (2.3) and lower bounds on $\varepsilon_{3,k}(x)$ from Lemma 9.0.64, we get

$$\frac{g_k(E)}{\binom{n}{k}} > \frac{c}{n}$$

for $3 \leq k \leq n-3$ when n is large enough.

In order to carry out the plan described above, we will need good lower bounds on $\varepsilon_2(x)$, $\varepsilon_{2,k}(x)$ and $\varepsilon_{3,k}(x)$ for $x \in S_n$. We will make use of the result known as Burnside's lemma (see for example [**DiMo**, Theorem 1.7A]).

LEMMA 2.0.7. *Let G be a finite group acting with ω orbits on a finite set X. For $g \in G$ let $\phi(g)$ be the number of elements of X fixed by g. Then*

$$\omega = \frac{1}{|G|}\sum_{g\in G}\phi(g).$$

2. NOTATION AND BASIC LEMMAS

DEFINITION 2.0.8. For $x \in S_n$ and $j, k \in [n]$, define
$$\delta_{j,k}(x) := \frac{f_j(x)}{\binom{n}{j}} - \frac{f_k(x)}{\binom{n}{k}}.$$

The next proposition follows immediately from Burnside's lemma.

PROPOSITION 2.0.9. For $x \in S_n$ and $j, k \in [n]$ we have
$$\varepsilon_{j,k}(x) = \frac{1}{|x|} \sum_{i=1}^{|x|} \delta_{j,k}(x^i).$$

Now we introduce some notation. Let $P(n)$ be the set of partitions of n. We will represent an element of $P(n)$ in either of two ways. Namely, $(\lambda_1, \ldots, \lambda_r) \in P(n)$ has parts $\lambda_1 \geq \ldots \geq \lambda_r$, while $1^{a_1} 2^{a_2} \ldots n^{a_n} \in P(n)$ has a_i parts equal to i. If $a_i = 0$ then i^{a_i} need not appear when the second representation is used. For $x \in S_n$ and $i \in [n]$, the number of cycles of length i in the decomposition of x as a product of disjoint cycles will be denoted by $c_i(x)$. We say $x \in S_n$ has cycle shape (or shape) $\pi \in P(n)$ if $\pi = 1^{c_1(x)} 2^{c_2(x)} \ldots n^{c_n(x)}$. We say $(\lambda_1, \ldots, \lambda_s)$ refines $(\mu_1, \ldots, \mu_t)$ if there is some partition $[s] = \bigcup_{j=1}^t A_j$ such that $\sum_{i \in A_j} \lambda_i = \mu_j$ for each $j \in [t]$.

Note that $x \in S_n$ fixes $Y \subseteq [n]$ if and only if Y is the union of some orbits of $\langle x \rangle$ on $[n]$. An immediate consequence of this fact is the next proposition, which we will use repeatedly without explicit reference.

PROPOSITION 2.0.10. Let $x, y \in S_n$. If the shape of x refines the shape of y then $f_k(x) \geq f_k(y)$ for all $k \in [n]$.

In the course of showing $g_2(E) > dn$ for most E, we will also show that $3g_2(E) > r$ for these E. However, since we don't know the constant d, we cannot use Lemma 2.0.6 to show that $3g_k(E) > r$ for all k between 2 and $n-2$ when $3g_2(E) > r$. We will obtain this result using the basic theory of covering spaces along with the representation theory of the symmetric group. The next lemma seems to be well known, a version which works in all characteristics appears in [**Gu1**]. The module V mentioned in the lemma is the first homology group of the quotient of a surface with monodromy data $(G, 1, E)$ by the action of H.

LEMMA 2.0.11. Let G be a finite group and let $H \leq G$. Let $\phi_H : G \to S_n$ be a homomorphism determined by the action of G on the cosets of H by translation. Fix an r-tuple $E = (x_1, \ldots, x_r)$ from G with $\prod_{i=1}^r x_i = 1$ and $\langle E \rangle = G$. There exists a $\mathbb{C}[G]$-module V such that every analytic map $f : X \to \mathbf{P}^1$ with monodromy data $(\phi_H(G), \phi_H(H), \phi_H(E))$ satisfies
$$2g(X) = \langle V, 1 \rangle_H.$$

LEMMA 2.0.12. Let $E = (x_1, \ldots, x_r) \in (S_n \setminus \{1\})^r$ with $A_n \leq \langle E \rangle$ and $\prod_{i=1}^r x_i = 1$. Then for $1 \leq k < \frac{n}{2}$ we have $g_k(E) \leq g_{k+1}(E)$.

PROOF. For $G \in \{S_n, A_n\}$ let V be the $\mathbb{C}[G]$-module described in Lemma 2.0.11. For any $H \leq G$, Frobenius reciprocity (see for example [**Is**, Lemma 5.2]) gives
$$\langle V, 1 \rangle_H = \langle V, 1_H^G \rangle_G.$$

Therefore, if $G(k) \leq G$ is the stabilizer of a k-set we have

(2.5) $$2g_k(E) = \langle V, 1_{G(k)}^G \rangle_G.$$

It is well known (see for example [**Sa**]) that the irreducible representations of S_n over $\mathbb{C}$ are parameterized by partitions of $[n]$ and that if $\mathcal{S}^\lambda$ denotes the representation corresponding to λ then for $0 \leq k \leq \lfloor \frac{n}{2} \rfloor$ we have

$$1_{S_n(k)}^{S_n} = \bigoplus_{j=0}^{k} \mathcal{S}^{(n-j,j)}.$$

Therefore, if $k < \lfloor \frac{n}{2} \rfloor$ then $1_{S_n(k)}^{S_n}$ is a submodule of $1_{S_n(k+1)}^{S_n}$. The lemma (for $G = S_n$) now follows from equation (2.5). It follows from Mackey's Lemma (see for example [**Is**, Problem 5.6]) that

$$1_{A_n(k)}^{A_n} = (1_{S_n(k)}^{S_n})_{A_n},$$

which proves the lemma for $G = A_n$. □

LEMMA 2.0.13. *Assume $n \geq 5$. Let $G \in \{A_n, S_n\}$ and let $H < G$ with $A_n \not\leq H$. Let $E = (x_1, \ldots, x_r)$ be an r-tuple of nonidentity elements of S_n with $\langle E \rangle = G$ and $\prod_{i=1}^r x_i = 1$. Then*

$$g(G, H, E) \geq \sum_{k=1}^{\lfloor \frac{n}{2} \rfloor} (o_k(H) - o_{k-1}(H))(g_k(E) - g_{k-1}(E)).$$

In particular, for $2 \leq k \leq \lfloor \frac{n}{2} \rfloor$, we have

$$g(G, H, E) \geq (o_k(H) - o_{k-1}(H))(g_k(E) - g_{k-1}(E)).$$

PROOF. We begin with the case $G = S_n$. Let V be the module from Lemma 2.0.11 which satisfies

$$2g(G, H, E) = \langle V, 1 \rangle_H = \langle V, 1_H^G \rangle_G.$$

We have

$$\langle V, 1_H^G \rangle_G = \sum_\lambda \langle 1_H^G, \mathcal{S}^\lambda \rangle \langle V, \mathcal{S}^\lambda \rangle,$$

the sum being over all partitions λ of n. In particular, we have

$$\langle V, 1_H^G \rangle_G \geq \sum_{k=1}^{\lfloor \frac{n}{2} \rfloor} \langle 1_H^G, \mathcal{S}^{(n-k,k)} \rangle \langle V, \mathcal{S}^{(n-k,k)} \rangle$$

$$= \sum_{k=1}^{\lfloor \frac{n}{2} \rfloor} \langle 1_H^G, 1_{G(k)}^G - 1_{G(k-1)}^G \rangle \langle V, 1_{G(k)}^G - 1_{G(k-1)}^G \rangle,$$

the equality $\mathcal{S}^{(n-k,k)} = 1_{G(k)}^G - 1_{G(k-1)}^G$ having been discussed in the proof of Lemma 2.0.12. Now, by Lemma 2.0.11, we have

$$\langle V, 1_{G(j)}^G \rangle = 2g_j(E)$$

for all j, and, by Frobenius reciprocity and Burnside's lemma, we have

$$\langle 1_H^G, 1_{G(j)}^G \rangle = \langle 1, 1_{G(j)}^G \rangle_H = o_j(H).$$

This completes the proof in the case $G = S_n$. Now the condition $n \geq 5$ guarantees that no partition $(n - k, k)$ is self-conjugate. It follows (see for example [**JaKe**]) that the restriction of each $\mathcal{S}^{(n-k,k)}$ to A_n is irreducible, and the proof just given for $G = S_n$ can be easily modified to work when $G = A_n$. □

The next result and its corollaries allow us to reduce the number of r-tuples E which must be examined when proving Theorems 1.1.2 and 1.2.1.

Let $c_1, \ldots, c_r$ be positive integers. Set

$$\gamma := \gamma(c_1, \ldots, c_r) := 2 - \sum_{i=1}^{r} \frac{c_i - 1}{c_i}$$

and assume that $\gamma > 0$. Set

$$m := m(c_1, \ldots, c_r) = \frac{2}{2 - \gamma}.$$

Thus $2 - \gamma = 2/m$.

Note that m is a positive integer. Indeed m is the order of the group generated by $x_i, 1 \leq i \leq r$ subject only to the relations that $x_i^{d_i} = 1$ and $x_1 \cdots x_r = 1$.

The following result is essentially in [**GuNe**] - see sections 3 and 4 of that paper. Note that it suffices to take G to be the free group on $r - 1$ generators (viewed as the group generated by $x_1, \ldots, x_r$ subject to the product being 1). Then N is a normal subgroup of index m. View G as the fundamental group of the Riemann sphere with r points removed. Consider the Riemann surface corresponding to N. Then N has finite index m in N and is the fundamental group of that Riemann surface and has the desired presentation. One can give a purely group theoretic proof as well (indeed, except for the $(2,3,5)$ case, it reduces to consider the case where G/N is cyclic).

LEMMA 2.0.14. *Suppose that $G = \langle x_1, \ldots, x_r \rangle$ with $\prod_{i=1}^{r} x_i = 1$. For $1 \leq i \leq r$ let c_i be a positive integer with $\gamma = \gamma(c_1, \ldots, c_r) > 0$. Let N be the normal closure in G of the set $\{x_1^{c_1}, \ldots, x_r^{c_r}\}$. Then N can be generated by m/c_i conjugates of $x_i^{c_i}$ with product 1 (in some order), where $m = m(c_1, \ldots, c_r)$.*

Suppose that G is generated by the r-tuple $E = (x_1, \ldots, x_r)$ with product 1 and G acts transitively on Ω of cardinality n. Let d_i be the order of x_i.

If $c|d$ are positive integers, let $J(c, d)$ be the set of all $j \in [d]$ which are not multiples of c. If x has order d and $c|d$, set

$$V(x, c) = \frac{1}{d} \sum_{j \in J(c,d)} f(x^j),$$

where $f(y)$ is the number of fixed points of y on Ω.

A straightforward computation (see [**GuNe**]) yields that if $c|d$ and x has order d and is a permutation on Ω of cardinality n, then

$$\mathrm{ind}(x, \Omega) = (c - 1)n/c - V(x, c) + (1/c)\mathrm{ind}(x^c).$$

Now choose positive integers $c_1, \ldots, c_r$ with $c_i|d_i$ and assume that

$$\sum_{i=1}^{r} \frac{c_i - 1}{c_i} < 2.$$

Let $m = m(c_1, \ldots, c_r)$.

Let N be the normal closure of $\langle x_1^{c_1}, \ldots x_r^{c_r}\rangle$. By the previous lemma, there is a generating r-tuple F of N consisting of m/c_i conjugates of $x_i, 1 \le i \le r$. Let $g = g(G, \Omega, E)$ and $h = g(N, \Omega, F)$ (this is well defined even if N is not transitive on Ω - if N is transitive, then $h \ge 0$).

LEMMA 2.0.15. *We have*
$$g - 1 = \frac{h-1}{m} - \frac{1}{2}\sum_{i=1}^{r} V(x_i, c_i).$$

PROOF. We have
$$2(n + g - 1) = \sum_{i=1}^{r} ind(x_i) = n\sum_{i=1}^{r} \frac{c_i - 1}{c_i} + \sum_{i=1}^{r} \frac{ind(x_i^{c_i})}{c_i} - \sum_{i=1}^{r}(V(x_i, c_i)).$$

This yields
$$2(g-1) = -(2-\gamma)n + \frac{1}{m}\sum_{y \in F} ind(y) - \sum_{i=1}^{r}(V(x_i, c_i)).$$

Since $\frac{1}{m}\sum_{y \in F} ind(y) = 2(n+h-1)$ and $(2-\gamma) = 2/m$, the result follows. □

We note some consequences. Let G, Ω, E be as above. The first result below was observed by P. Müller in [**Mu**].

PROPOSITION 2.0.16. *Suppose that $g = 0$ and all orbits of x_1, x_2 have length divisible by $c > 1$. If N is the normal closure in G of $\langle x_1^c, x_2^c, x_3, \ldots, x_r\rangle$, then N is not transitive. In particular, if G is primitive, then $N = 1$ and G is cyclic of prime order c.*

PROOF. If G is primitive, then any nontrivial normal subgroup is transitive. Thus, the last statement follows immediately.

Let $c_1 = c_2 = c$ and $c_i = 1$ for $i > 2$. The condition that all orbits of x_i have size a multiple of c for $i = 1, 2$ implies that $V(x_i, c_i) = 0$. Applying the previous lemma yields that $h < 0$, whence N is not transitive. □

COROLLARY 2.0.17. *Suppose that $g = 0$ and the sum of the lengths of all orbits of odd length of x_1, x_2 is 2. Assume that all orbits of x_3 have length divisible by $c > 1$. If N is the normal subgroup of G generated by $x_1^2, x_2^2, x_3^c, x_4, \ldots, x_r$, then N is not transitive. In particular, if G is primitive, then $N = 1$ and G is dihedral.*

PROOF. As above, the last statement follows immediately from the fact that a nontrivial normal subgroup of a primitive permutation group is transitive.

Let $c_1 = c_2 = 2, c_3 = c$ and $c_i = 1$ for $i > 3$. The hypotheses imply that $V(x_i, c_i) = 0$ for $i > 2$ and $V(x_1, 2) + V(x_2, 2) \le 1$. Applying Lemma 2.0.15 yields that
$$\frac{h}{m} = -1 + \frac{1}{m} + \frac{1}{2}\sum_{i=1}^{r} V(x_i, c_i) \le \frac{1}{2} + \frac{1}{m}.$$
Since $m = 2c > 4$, this implies that $h < 0$, whence N is not transitive. □

COROLLARY 2.0.18. *Suppose that $g = 0$, $(d_1, d_2, d_3) = (2, 3, 5)$. Moreover, assume that $\sum_{i=1}^{3} f(x_i) \le 2$. If N is the normal subgroup of G generated by $x_1^2, x_2^3, x_3^5, \ldots, x_r$, then N is not transitive. In particular, if G is primitive, then $N = 1$ and $G \cong A_5$.*

PROOF. In this case $m = 60$. Applying the lemma, we get
$$\frac{h}{60} = -1 + \frac{1}{60} + \frac{1}{2}\sum V(x_i, c_i),$$
where $c_1 = 2, c_2 = 3, c_3 = 5$ and $c_j = 1$ for $j > 3$. The hypotheses imply that $\sum V(x_i, c_i) \leq \frac{8}{5}$, so $\frac{h}{60} \leq -\frac{59}{60} + \frac{4}{5} < 0$. Thus, $h < 0$ and N is not transitive. If G is primitive, then $N = 1$, whence $G = A_5$. □

A *transposition* in S_n is an element of shape $1^{n-2}2^1$. Transpositions play a large role in each of our main theorems, and the next lemma plays an important role in each of their proofs.

LEMMA 2.0.19. *Let n, k be integers with $0 < k < n$. Let $t \in S_n$ be a transposition on $[n]$. Then for any nonidentity $x \in S_n$ we have*
$$i_k(x) \geq i_k(t) = \binom{n-2}{k-1}.$$

PROOF. We assume that $t = (12)$. We first prove the inequality by showing that $o_k(x) \leq o_k(t)$. If $x \neq 1$ then we may assume that 1 and 2 are in the same cycle of x. Then every k-set which is fixed by x is also fixed by t. Since every orbit of $\langle t \rangle$ on $\binom{[n]}{k}$ has size at most two, we see that for each element $X \in \binom{[n]}{k}$ the $\langle x \rangle$-orbit containing X is at least as large as the $\langle t \rangle$-orbit containing X. It follows that $o_k(t) \geq o_k(x)$, so $i_k(t) \leq i_k(x)$.

Let $X \in \binom{[n]}{k}$. If $|X \cap \{1, 2\}| \neq 1$ then X is fixed by t, while if $|X \cap \{1, 2\}| = 1$ then X is fixed by t^2 but not by t. It follows that $o_k(t) = \binom{n}{k} - \binom{n-2}{k-1}$. □

Throughout the paper, we treat (ordered) r-tuples of elements of S_n as if they are (unordered) multisets, often writing $\{x_1, \ldots, x_r\}$ rather than $(x_1, \ldots, x_r)$. The next lemma, which is well known and appears in [**Fri1**] says that this assumption is harmless.

LEMMA 2.0.20. *Let $x_1, \ldots, x_r \in S_n$ with $\prod_{i=1}^r x_i = 1$. For each $\sigma \in S_r$ there exist $y_1, \ldots, y_r \in S_n$ such that*
- $\langle y_1, \ldots, y_r \rangle = \langle x_1, \ldots, x_r \rangle$,
- $\prod_{i=1}^r y_i = 1$, *and*
- *for each $i \in [r]$, y_i is conjugate to $x_{i\sigma}$.*

CHAPTER 3

Examples

In this chapter, we exhibit several infinite classes of examples of triples (G, H, E) such that $G \in \{S_n, A_n\}$, H is the stabilizer of a 2-set in G and $g(G, H, E) = 0$. These classes include those mentioned in Theorems 1.1.1 and 1.2.1. Many of these examples are obtained by appropriately modifying examples where $E = (t, z, y)$, with t a transposition, z an n-cycle and (necessarily) y of shape $(a, n-a)$ with $\gcd(n, a) = 1$. Such examples are treated in detail in Chapter 10. Let us record here the main result of that chapter, Theorem 10.0.66.

THEOREM 3.0.21. *Assume that $t \in S_n$ has shape $1^{n-2}2^1$, y has shape $a^1 b^1$ with $\gcd(a, n) = 1$, z has shape n^1 and $tzy = 1$. Let $E = (t, z, y)$. Then*
   (1) *We have $g_1(E) = g_2(E) = 0$.*
   (2) *There exists some constant $c > 0$ such that if $n > 7$ and $3 \le k \le n-3$ then*
$$\frac{g_k(E)}{\binom{n}{k}} > \frac{c}{n}$$
   (3) *If $n > 6$ then $g_3(E) > 2$ unless one of the conditions in the table below holds.*

n	a	$g_3(E)$
7	6	0
7	5	1
7	4	1
8	7	1
8	5	2
9	8	1
9	7	2
10	9	2

   (4) *If $n > 9$ then $g_4(E) - g_3(E) > 2$.*

Note that the condition $\gcd(a, n) = 1$ guarantees that the (obviously transitive) group $\langle t, z \rangle$ is primitive, and the presence of the transposition t then guarantees that $\langle t, z \rangle = S_n$. (One can also prove this last fact by elementary methods.) The next two lemmas allow us to make the modifications mentioned above.

LEMMA 3.0.22. *Let $z \in S_n$ be an n-cycle and let $u, v \in S_n$ be involutions with $uv = z$. If $k \le 2$ then*
$$i_k(z) = i_k(u) + i_k(v).$$

PROOF. If n is even we may assume that u has shape $2^{n/2}$ and v has shape $1^2 2^{(n-2)/2}$ while if n is odd then both u and v have shape $1^1 2^{(n-1)/2}$. (This follows easily from the fact that the dihedral group generated by u and v contains z.) The

claim of the lemma in the case $k = 1$ follows immediately. Using Burnside's Lemma, we calculate that
$$i_2(z) = \begin{cases} \frac{n^2-2n+1}{2}, & n \text{ odd} \\ \frac{n^2-2n}{2}, & n \text{ even}, \end{cases}$$
while
$$i_2(u) = i_2(v) = \begin{cases} \frac{n^2-2n+1}{4}, & n \text{ odd} \\ \frac{n^2-2n}{4}, & n \text{ even}. \end{cases}$$
The lemma follows immediately. □

LEMMA 3.0.23. *Let $y \in S_n$ have shape $(a, n-a)$ with $\gcd(a, n) = 1$. Let $u, v \in S_n$ be involutions with $uv = y$. If $k \le 2$ then*
$$i_k(y) = i_k(u) + i_k(v).$$

PROOF. If n is even then both a and $n - a$ are odd and both of u, v have cycle shape $1^2 2^{(n-2)/2}$, while if n is odd then one of $a, n-a$ is even and we may assume that u has cycle shape $1^1 2^{(n-1)/2}$ and v has cycle shape $1^3 2^{(n-3)/2}$. (To see this, note that both u and v normalize $\langle y \rangle$, and since y has two orbits of distinct size, both u and v preserve both orbits. We can now argue as in the proof of Lemma 3.0.22.) The claim of the lemma when $k = 1$ follows immediately. As shown in the proof of Theorem 10.0.66, Burnside's Lemma gives
$$i_2(y) = \begin{cases} \frac{n^2-2n}{2}, & n \text{ even} \\ \frac{n^2-2n-1}{2} & n \text{ odd}. \end{cases}$$
When $w \in \{u, v\}$ has one or two fixed points, $i_2(w)$ is as calculated in the proof of Lemma 3.0.22. If n is odd and v has three fixed points, Burnside's Lemma gives
$$i_2(v) = \frac{n^2 - 2n - 3}{4}.$$
In any case, our lemma now follows from direct calculation. □

Say $E = (x_1, \ldots, x_r)$ satisfies $\prod_{i=1}^r x_i = 1$ and $A_n \le \langle E \rangle$. If $u, v \in S_n$ with $uv = x_i$ then $E' := (x_1, \ldots, x_{i-1}, u, v, x_{i+1}, \ldots, x_r)$ satisfies the same two properties. Thus the next result follows immediately from the Riemann-Hurwitz formula and the results just given.

PROPOSITION 3.0.24. *For each of the conditions (a)-(f) below and each n of appropriate parity, there exist r-tuples $E = (x_1, \ldots, x_r)$ from S_n with $\prod_{i=1}^r x_i = 1$, $\langle E \rangle = S_n$ and $g_2(E) = 0$ satisfying the given condition.*
  (a) *E consists of one element of cycle shape $1^{n-2}2^1$, one of shape $2^{\frac{n}{2}}$ and three of shape $1^2 2^{\frac{n-2}{2}}$.*
  (b) *E consists of one element of cycle shape $1^{n-2}2^1$, one of shape $1^3 2^{\frac{n-3}{2}}$ and three of shape $1^1 2^{\frac{n-1}{2}}$.*
  (c) *E consists of one element of cycle shape $1^{n-2}2^1$, two of shape $1^2 2^{\frac{n-2}{2}}$ and one of shape n^1.*
  (d) *E consists of one element of cycle shape $1^{n-2}2^1$, one of shape $1^2 2^{\frac{n-2}{2}}$, one of shape $2^{\frac{n}{2}}$ and one of shape $a^1 b^1$ with $\gcd(a, n) = 1$.*
  (e) *E consists of one element of cycle shape $1^{n-2}2^1$, one of shape $1^3 2^{\frac{n-3}{2}}$, one of shape $1^1 2^{\frac{n-1}{2}}$ and one of shape n^1.*

(f) E consists of one element of cycle shape $1^{n-2}2^1$, two of shape $1^1 2^{\frac{n-1}{2}}$ and one of shape $a^1 b^1$ with $\gcd(a,n)=1$.

In Chapter 12, we examine the case where $E = (x_1, \ldots, x_r)$ and x_r is an n-cycle. There are several classes of examples where $g_2(E) = 0$ besides the one given in Theorem 10.0.66. These examples are obtained by starting with involutions u, v such that $z^{-1} = uv$ is an n-cycle, finding some involution $t \in S_n$ such that $A_n \leq \langle ut, tv \rangle$ and then taking $E = (ut, tv, z)$. We describe them here.

PROPOSITION 3.0.25. *Assume $n \geq 7$. Then for each of the conditions (a)-(f) below and each n of appropriate parity, there exist triples $E = (x_1, x_2, x_3)$ from S_n satisfying the given condition, such that $A_n \leq \langle E \rangle$, $x_1 x_2 x_3 = 1$, $g_1(E) = g_2(E) = 0$ and x_3 is an n-cycle.*

(a) *The element x_1 has cycle shape $1^3 2^{\frac{n-3}{2}}$ and x_2 has shape $1^1 2^{\frac{n-5}{2}} 4^1$.*
(b) *The element x_1 has cycle shape $1^3 2^{\frac{n-3}{2}}$ and x_2 has shape $2^{\frac{n-3}{2}} 3^1$.*
(c) *The element x_1 has cycle shape $1^2 2^{\frac{n-2}{2}}$ and x_2 has shape $1^2 2^{\frac{n-6}{2}} 4^1$.*
(d) *The element x_1 has cycle shape $1^2 2^{\frac{n-2}{2}}$ and x_2 has shape $1^1 2^{\frac{n-4}{2}} 3^1$.*
(e) *The element x_1 has cycle shape $1^1 2^{\frac{n-1}{2}}$ and x_2 has shape $1^3 2^{\frac{n-7}{2}} 4^1$.*
(f) *The element x_1 has cycle shape $1^1 2^{\frac{n-1}{2}}$ and x_2 has shape $1^2 2^{\frac{n-5}{2}} 3^1$.*

PROOF. In each case, direct computations show that $g_1(E) = g_2(E) = 0$, so we only need to prove the existence of systems of the types listed in the theorem. Assume first that n is odd. Set

$$u := \prod_{i=1}^{\frac{n-1}{2}} (i, n-i),$$

and

$$v := \prod_{i=2}^{\frac{n+1}{2}} (i, n+2-i).$$

Then $\langle u, v \rangle$ is a dihedral group of order $2n$ and $uv = (1, 2, \ldots, n)^2$ is an n-cycle. Let $z = (uv)^{-1}$. To get a system $E = \{x, y, z\}$ which satisfies condition (a), we set

$$t := (2, n-2)$$

and let $x = ut$, $y = tv$. Then

$$xyz = rt^2 sz = rsz = 1.$$

Also,

$$x = (1, n-1) \prod_{i=3}^{\frac{n-1}{2}} (i, n-i)$$

has shape $1^3 2^{\frac{n-3}{2}}$ and

$$y = (2, 4, n-2, n)(3, n-1) \prod_{i=5}^{\frac{n+1}{2}} (i, n+2-i)$$

has shape $1^1 2^{\frac{n-5}{2}} 4^1$. It remains to show that $A_n \leq G := \langle x, y, z \rangle$. Certainly G is transitive, as it contains z. Since $\langle z \rangle$ contains $(1, 2, \ldots, n)$, if $\Pi = [\pi_1 | \ldots | \pi_d]$ is a G-invariant partition of $[n]$ then $d | n$ and we may assume that

$$\pi_i = \{j \in [n] : j \equiv i \bmod d\}$$

for each $i \in [d]$. Now $y^2 = (2, n-2)(4, n) \in G$, and since n is odd y^2 does not stabilize any partition of the type described above. Therefore G is primitive. Applying a theorem of Jordan (see [**DiMo**, Example 3.3.1]), to the element y^2, we see that $A_n \leq G$. The same construction using $t = (1, n-1)$, $t = (\frac{n-3}{2}, \frac{n-1}{2})(\frac{n+1}{2}, \frac{n+3}{2})$ and $t = (1, n-3)(3, n-1)$ yields systems satisfying conditions (b),(e) and (f), respectively. In each case we get elements $x = ut$, $y = tv$ of the desired shapes such that y^2 does not fix any partition of $[n]$ which is fixed by z and has shape $1^{n-4}2^2$ or $1^{n-3}3^1$. It follows from either the theorem of Jordan cited above or [**DiMo**, Theorem 3.3A] that $A_n \leq \langle x, y, z \rangle$.

Now assume that n is even. Set

$$u := \prod_{i=2}^{\frac{n}{2}} (i, n+2-i)$$

and

$$v := \prod_{i=1}^{\frac{n}{2}} (i, n+1-i).$$

To get a system $E = \{x, y, z\}$ satisfying condition (d), we set $t = (1n)$ and let $x = ut$, $y = tv$ as above. To get a system satisfying condition (c) in this manner, we must be a little more careful in order to guarantee that the group $\langle x, y, z \rangle$ is primitive. Let $p < \frac{n}{2}$ be a prime which does not divide n (such a p exists since $n > 6$) and let $a = \frac{p+1}{2}$, so $1 < a < \frac{n}{2}$. Set

$$t = (a, n+1-a)$$

and let $x = ut$, $y = tv$ as usual. Then

$$y = \prod_{i \neq a} (i, n+1-i)$$

has shape $1^2 2^{\frac{n-2}{2}}$ and, if we set $X := \{1, a, a+1, \frac{n}{2}+1, n+1-a, n+2-a\}$ then

$$x = (a, a+1, n+1-a, n+2-a) \prod_{i \notin X} (i, n+2-i)$$

has shape $1^2 2^{\frac{n-6}{2}} 4^1$. Also, $x^2 = (a, n+1-a)(a+1, n+2-a)$ and the facts that $(n+1-a) - a = n - p$ is relatively prime to n and $a < \frac{n+2}{4}$ together guarantee that x^2 does not fix any nontrivial partition of $[n]$ which is fixed by z. Therefore $A_n \leq \langle x, y, z \rangle$ as above. □

Applying Lemma 3.0.22 to each of the examples given in Proposition 3.0.25, we get the following classes of examples.

PROPOSITION 3.0.26. *For each of the conditions (a)-(f) below and each $n \geq 7$ of appropriate parity, there exist 4-tuples $E = (x_1, x_2, x_3, x_4)$ from S_n satisfying the given condition, such that $A_n \leq \langle E \rangle$, $x_1 x_2 x_3 x_4 = 1$ and $g_1(E) = g_2(E) = 0$.*

(a) *E contains two elements of cycle shape $1^1 2^{(n-1)/2}$, one of shape $1^3 2^{\frac{n-3}{2}}$ and one of shape $1^1 2^{\frac{n-5}{2}} 4^1$.*
(b) *E contains two elements of cycle shape $1^1 2^{(n-1)/2}$, one of shape $1^3 2^{\frac{n-3}{2}}$ and one of shape $2^{\frac{n-3}{2}} 3^1$.*

(c) E contains two elements of cycle shape $1^2 2^{\frac{n-2}{2}}$, one of shape $2^{\frac{n}{2}}$ and one of shape $1^2 2^{\frac{n-6}{2}} 4^1$.
(d) E contains two elements of cycle shape $1^2 2^{\frac{n-2}{2}}$, one of shape $2^{\frac{n}{2}}$ and one of shape $1^1 2^{\frac{n-4}{2}} 3^1$.
(e) E contains three elements of cycle shape $1^1 2^{\frac{n-1}{2}}$ and one of shape $1^3 2^{\frac{n-7}{2}} 4^1$.
(f) E contains three elements of cycle shape $1^1 2^{\frac{n-1}{2}}$ and one of shape $1^2 2^{\frac{n-5}{2}} 3^1$.

Applying Lemma 3.0.23 to the examples described in Proposition 3.0.26(a,b,c,d), we get the following classes of examples.

PROPOSITION 3.0.27. *Assume $a + b = n \geq 7$ and $\gcd(a,b) = 1$. For each of the conditions (a)-(d) below, if n has the appropriate parity then there exist triples $E = (x_1, x_2, x_3)$ from S_n, satisfying the given condition, such that $A_n \leq \langle E \rangle$, $x_1 x_2 x_3 = 1$, $g_1(E) = g_2(E) = 0$ and x_3 has cycle shape $a^1 b^1$.*

(a) *The element x_1 has cycle shape $1^1 2^{\frac{n-1}{2}}$ and x_2 has shape $1^1 2^{\frac{n-5}{2}} 4^1$.*
(b) *The element x_1 has cycle shape $1^1 2^{\frac{n-1}{2}}$ and x_2 has shape $2^{\frac{n-3}{2}} 3^1$.*
(c) *The element x_1 has cycle shape $2^{\frac{n}{2}}$ and x_2 has shape $1^2 2^{\frac{n-6}{2}} 4^1$.*
(d) *The element x_1 has cycle shape $2^{\frac{n}{2}}$ and x_2 has shape $1^1 2^{\frac{n-4}{2}} 3^1$.*

Next we present some infinite classes of triples E such that $g_1(E) = g_2(E) = 0$ which are obtained by adjusting triples which generate a dihedral group.

PROPOSITION 3.0.28. *Assume $a + b = n$ and $\gcd(a,b) = 1$. For each of the conditions (a)-(e) below, if n has the appropriate parity then there exist triples $E = (x_1, x_2, x_3)$ from S_n satisfying the given condition, such that $A_n \leq \langle E \rangle$, $x_1 x_2 x_3 = 1$, $g_1(E) = g_2(E) = 0$ and x_3 has cycle shape $a^1 b^1$.*

(a) *The element x_1 has cycle shape $2^{\frac{n}{2}}$ and x_2 has shape $1^2 2^{\frac{n-6}{2}} 4^1$.*
(b) *The element x_1 has cycle shape $2^{\frac{n}{2}}$ and x_2 has shape $1^1 2^{\frac{n-4}{2}} 3^1$.*
(c) *The element x_1 has cycle shape $1^2 2^{\frac{n-2}{2}}$ and x_2 has shape $2^{\frac{n-4}{2}} 4^1$.*
(d) *The element x_1 has cycle shape $1^1 2^{\frac{n-1}{2}}$ and x_2 has shape $1^1 2^{\frac{n-5}{2}} 4^1$.*
(e) *The element x_1 has cycle shape $1^1 2^{\frac{n-1}{2}}$ and x_2 has shape $2^{\frac{n-3}{2}} 3^1$.*

PROOF. We prove first that it suffices to describe triples E satisfying the conditions (a)-(e) with $x_1 x_2 x_3 = 1$ and $\langle E \rangle$ transitive. Indeed, if $\langle E \rangle$ is transitive then it must be primitive, since (assuming $a < b$) its element x_3^a has cycle shape $1^a b^1$ and therefore cannot stabilize any proper nontrivial partition of $[n]$. (If it did, it would permute transitively a set of parts whose union has size b, but $\gcd(b, n) = 1$.) Moreover, x_2^2 has shape $1^{n-4} 2^2$ or $1^{n-3} 3^1$, and in either case we get $A_n \leq \langle E \rangle$ ([**DiMo**, Theorem 3.3A]). Finally, direct calculation will give $g_1(E) = g_2(E) = 0$.

To produce the desired triples for conditions (b) and (e), we begin by identifying $[n]$ with $\mathbb{Z}_n$ in the natural manner, and letting w, x be the involutions mapping $j \in \mathbb{Z}_n$ to $-j, 1-j$, respectively. Thus w fixes only n and $\frac{n}{2}$ if n is even and fixes only n if n is odd, while x is fixed-point-free if n is even and fixes only $\frac{n+1}{2}$ if n is odd. For all n, we see that $z^{-1} := xw$ maps j to $j-1$ for all $j \in \mathbb{Z}_n$, and is therefore an n-cycle. For any $a \in \mathbb{Z}_n$, let t_a be the transposition exchanging a and n while fixing all other points. Set $x_1 = x$, $x_2 = w t_a$ and $x_3 = t_a z$. Since $xwz = t_a^2 = 1$, we have $x_1 x_2 x_3 = 1$ as desired. For $j \notin \{n, n-a\}$, we have $j x_2 = n - j$, while $(n-a) x_2 = n$ and $n x_2 = a$. For $j \notin \{a, n\}$, we have $j x_3 = j+1$, while $a x_3 = 1$ and

$nx_3 = a+1$. Thus x_1, x_2, x_3 have the desired cycle shapes. Since $\langle x_3 \rangle$ has orbits $\{1, \ldots, a\}$ and $\{a+1, \ldots, n\}$ and $nx_2 = a$, we see that $\langle E \rangle$ is transitive as needed.

For condition (c), let w, x, z be as above. For any $a \in \mathbb{Z}_n$, let s_a be the transposition exchanging 1 and a while fixing all other points. Set $x_1 = w$, $x_2 = xs_{a+1}$ and $x_3 = s_{a+1}z^{-1}$. As above, we see that $x_1 x_2 x_3 = 1$. For $j \notin \{n-a, n\}$ we have $jx_2 = n+1-j$, while $nx_2 = a+1$ and $(n-a)x_2 = 1$. For $j \notin \{1, a+1\}$, we have $jx_3 = j-1$, while $1x_3 = a$ and $(a+1)x_3 = n$. Thus x_1, x_2, x_3 have the desired cycle shapes. The orbits of $\langle x_3 \rangle$ are $\{1, \ldots, a\}$ and $\{a+1, \ldots, n\}$. Since $1x_2 = n$, we see that $\langle E \rangle$ is transitive.

For conditions (a) and (d), let $u, v \in S_n$ be involutions satisfying

$$ju = \begin{cases} a - j & j \in \{1, \ldots, a-1\}, \\ a & j = a, \\ n + a - j & j \in \{a+1, \ldots, n-1\}, \\ n & j = n \end{cases}$$

and

$$jv = \begin{cases} a + 1 - j & j \in \{1, \ldots, a\}, \\ n + a + 1 - j & j \in \{a+1, \ldots, n\}. \end{cases}$$

If n is even, we know that a is odd. Thus u fixes only a and n, while v fixes only $\frac{a+1}{2}$ and $\frac{n+a+1}{2}$. If n is odd, assume that a is even. Then u fixes only $\frac{a}{2}$, a and n while v fixes only $\frac{n+a+1}{2}$. Regardless of the parity of n, uv maps j to $j+1$ for $j \notin \{a, n\}$, while $auv = 1$ and $nuv = a+1$. Thus uv has cycle shape $a^1 b^1$. With t_a as above, set $x_1 = ut_a$, $x_2 = t_a v$ and $x_3 = vu$. Certainly x_1 and x_3 have the desired cycle shapes, while

$$jx_2 = \begin{cases} a + 1 - j & j \in \{1, \ldots, a-1\}, \\ a + 1 & j = a, \\ n + a + 1 - j & j \in \{a+1, \ldots, n-1\}, \\ 1 & j = n. \end{cases}$$

Thus x_2 also has the desired cycle shape. Since $\langle x_3 \rangle$ has orbits $\{1, \ldots, a\}$ and $\{a+1, \ldots, n\}$ and $nx_2 = 1$, we see that $\langle E \rangle$ is transitive. □

Finally, computer calculations by the first author and R. Stafford indicate the existence of the following infinite families of examples.

CONJECTURE 3.0.29. *For each of the conditions (a)-(j) below and each n satisfying the appropriate modular equation, there exist triples $E = (x_1, x_2, x_3)$ from S_n satisfying the given condition, such that $A_n \leq \langle E \rangle$, $x_1 x_2 x_3 = 1$ and $g_1(E) = g_2(E) = 0$.*

(a) *E consists of one element of cycle shape $1^2 2^{\frac{n-2}{2}}$, one of shape $1^1 3^1 4^{\frac{n-4}{4}}$ and one of shape $4^{\frac{n}{4}}$.*

(b) *E consists of one element of cycle shape $1^1 2^{\frac{n-1}{2}}$, one of shape $1^1 4^{\frac{n-1}{4}}$ and one of shape $2^1 3^1 4^{\frac{n-5}{4}}$.*

(c) *E consists of one element of cycle shape $1^2 2^{\frac{n-2}{2}}$, one of shape $1^1 2^1 4^{\frac{n-3}{4}}$ and one of shape $3^1 4^{\frac{n-3}{4}}$.*

(d) *E consists of one element of cycle shape $1^2 2^{\frac{n-2}{2}}$, one of shape $1^1 2^1 3^{\frac{n-3}{3}}$ and one of shape $6^{\frac{n}{6}}$.*

(e) *E consists of one element of cycle shape $1^1 2^{\frac{n-1}{2}}$, one of shape $1^1 3^{\frac{n-1}{3}}$ and one of shape $3^1 4^1 6^{\frac{n-7}{6}}$.*

(f) *E consists of one element of cycle shape $2^{\frac{n}{2}}$, one of shape $2^1 3^{\frac{n-2}{3}}$ and one of shape $1^2 6^{\frac{n-2}{6}}$.*

(g) *E consists of one element of cycle shape $2^{\frac{n}{2}}$, one of shape $2^1 3^{\frac{n-2}{4}}$ and one of shape $2^1 6^{\frac{n-2}{6}}$.*

(h) *E consists of one element of cycle shape $1^1 2^{\frac{n-1}{2}}$, one of shape $1^1 2^1 3^{\frac{n-1}{3}}$ and one of shape $3^1 6^{\frac{n-3}{6}}$.*

(i) *E consists of one element of cycle shape $1^2 2^{\frac{n-2}{2}}$, one of shape $1^1 3^{\frac{n-3}{3}}$ and one of shape $4^1 6^{\frac{n-4}{6}}$.*

(j) *E consists of one element of cycle shape $1^1 2^{\frac{n-1}{2}}$, one of shape $2^1 3^{\frac{n-2}{3}}$ and one of shape $2^1 3^1 6^{\frac{n-5}{6}}$.*

CHAPTER 4

Proving the main results on five or more branch points - Theorems 1.1.1 and 1.1.2

In this Chapter, using results whose proofs appear in subsequent chapters, we prove Theorems 1.1.1 and 1.1.2. The key ingredient, and that whose proof requires the most work, is Theorem 1.1.1 under the additional assumption that $n > 20$, which is proved in Chapter 5. We restate this result here.

THEOREM 4.0.30. *Let $E = (x_1, \ldots, x_r)$ be an r-tuple from $S_n \setminus \{1\}$ with $r \geq 5$ and $n > 20$. Assume $A_n \leq \langle x_1, \ldots, x_r \rangle$ and $\prod_{i=1}^r x_i = 1$. Then there exists some constant $c > 0$ such that exactly one of the following conditions holds.*

(1) *We have $g_2(E) - g_1(E) > \max\{cn, 2\}$, and $g_2(E) - \frac{r}{3} > 0$.*
(2) *We have $g_2(E) = 0$ and one of the following two conditions holds.*
  (a) *The five elements of E have cycle shapes $1^{n-2}2^1$, $1^3 2^{\frac{n-3}{2}}$, $1^1 2^{\frac{n-1}{2}}$, $1^1 2^{\frac{n-1}{2}}$, $1^1 2^{\frac{n-1}{2}}$.*
  (b) *The five elements of E have cycle shapes $1^{n-2}2^1$, $1^2 2^{\frac{n-2}{2}}$, $1^2 2^{\frac{n-2}{2}}$, $1^2 2^{\frac{n-2}{2}}$, $2^{\frac{n}{2}}$.*

Moreover, for each n there exists some E which satisfies our assumptions and is described in condition 2.

We will also need the following result, which is proved in Chapter 6. The proof is similar to that of Theorem 4.0.30, although significantly easier.

THEOREM 4.0.31. *Let $E = (x_1, \ldots, x_r)$ be an r-tuple from $S_n \setminus \{1\}$ with $r \geq 5$ and $n > 20$. Assume $A_n \leq \langle x_1, \ldots, x_r \rangle$ and $\prod_{i=1}^r x_i = 1$. Then there exists some constant c' such that*

$$g_3(E) - g_2(E) \geq \max\{\frac{n-10}{3}, c'n^2\}.$$

In particular, $g_3(E) - g_2(E) > 2$.

We can now prove Theorem 1.1.2 for $n > 20$ as follows. Let H be a maximal subgroup of $G \in \{A_n, S_n\}$. If H is not transitive then H is the stabilizer in G of a k-set. The claim of the theorem now follows from Lemma 2.0.12 and Theorems 4.0.30 and 4.0.31. Assume now that H is transitive. By Lemma 2.0.13 and Theorem 4.0.30, if $o_2(H) > o_1(H)$ and E is not as described in Condition (2) of Theorem 4.0.30 then $g(G, H, E) > \max\{cn, 2\}$. Also, if $o_3(H) > o_2(H)$ then Lemma 2.0.13 and Theorem 4.0.31 allow us to conclude that $g(G, H, E) \geq \max\{c'n^2, 2\}$. Say H is imprimitive. Then H is the stabilizer in G of a partition of $[n]$ into k parts of size l, where $kl = n$ and $1 < k < n$. It is straightforward to show that

$$o_2(H) = 2$$

and
$$o_3(H) = \begin{cases} 2 & k \in \{2, \frac{n}{2}\}, \\ 3 & \text{otherwise.} \end{cases}$$
The claim of Theorem 1.1.2 follows in all cases except those where $k \in \{2, \frac{n}{2}\}$ and E is as described in Condition (2b) of Theorem 4.0.30. As noted in the proof of Theorem 4.0.31, in this case we have
$$g_3(E) = \frac{(n-4)^2}{4}.$$
Also, if $x \in S_n$ has shape $1^2 2^{(n-2)/2}$ or $2^{n/2}$ then
$$i_4(x) = \frac{1}{2}\left[\binom{n}{4} - \binom{n/2}{2}\right].$$
Combining this fact with Lemma 2.0.19 and direct calculation gives
$$g_4(E) = \frac{(n-4)(n-6)(2n-1)}{24}.$$
This gives
$$g_4(E) - g_3(E) = \frac{(n-2)(n-4)(2n-15)}{24} > 2.$$
It is straightforward to show that for either value of k we have
$$o_4(H) = 3,$$
and the claim of Theorem 1.1.2 now follows from Lemma 2.0.13.

Say H is primitive. We have the following result of P. Cameron, P. Neumann and J. Saxl ([**CaNeSa**]).

LEMMA 4.0.32. *Let $H \leq S_n$ be primitive. If $o_3(H) = o_2(H)$ then $o_3(H) = 1$.*

Thus the claim of Theorem 1.1.2 follows from Lemma 2.0.13 and Theorem 4.0.31 whenever H is not 3-homogeneous, and Theorem 1.1.2 (for $n > 20$) now follows from the next Theorem, which is proved in chapter 8.

THEOREM 4.0.33. *Let $G \in \{A_n, S_n\}$. Let $E = (x_1, \ldots, x_r)$ be an r-tuple of nonidentity elements of G such that $G = \langle E \rangle$ and $\prod_{i=1}^r x_i = 1$. Let $H \neq A_n$ be a 3-homogeneous maximal subgroup of G. Assume $r \geq 5$. Then there is some $c > 0$, which does not depend on G or H, such that one of the following conditions holds.*
  (1) *We have $g(G, H, E) > \max\{2, \frac{r}{3}, c[G:H]\}$.*
  (2) *$G = A_6$, $H = L_2(5)$ and $r \leq 9$.*
  (3) *$G = S_6$, $H = PGL_2(5)$ and $r \leq 14$.*
  (4) *$G = A_8$, $H = L_2(7)$ and E consists of five elements of cycle shape 2^4.*
  (5) *$G = A_8$, $H = AGL_3(2)$ and $r \leq 8$.*

The proof of Theorem 1.1.2 for $n \leq 20$ goes as follows. First we prove in Chapter 7 the following analogue of Theorem 4.0.30 under the assumption that $r \geq 9$.

THEOREM 4.0.34. *Let $E = (x_1, \ldots, x_r)$ be an r-tuple of nonidentity elements of S_n with $r \geq 9$ and $n \geq 5$. Assume $A_n \leq \langle x_1, \ldots, x_r \rangle$ and $\prod_{i=1}^r x_i = 1$. Then $g_2(E) - g_1(E) > 2$ and $g_2(E) - \frac{r}{3} > 0$.*

Now, arguing as we did in the proof for $n > 20$, we see that if $r \geq 9$ then $g(G, H, E) > 2$ whenever H is not 2-homogeneous. The proof for $r \geq 9$ is completed using the following result, whose proof appears in Chapter 8.

PROPOSITION 4.0.35. *Let $G \in \{A_n, S_n\}$. Let $E = (x_1, \ldots, x_r)$ be an r-tuple of nonidentity elements of G such that $G = \langle E \rangle$ and $\prod_{i=1}^{r} x_i = 1$. Let $H \neq A_n$ be a 2-homogeneous maximal subgroup of G. Assume $n \leq 20$ and $r \geq 5$. Then one of the following conditions holds.*

(1) *We have $g(G, H, E) > \max\{2, \frac{r}{3}\}$.*
(2) *One of conditions (2),(3),(4) or (5) of Theorem 4.0.33 holds.*
(3) *$G = A_7$, $H = L_2(7)$ (acting on cosets of S_4) and $r = 5$.*

There are now finitely many cases left to examine in order to finish the proofs of both theorems, namely, those where $5 \leq n \leq 20$ and $5 \leq r \leq 8$. These cases are handled by direct computer calculation, as described in Appendix A.

CHAPTER 5

Actions on 2-sets - the proof of Theorem 4.0.30

In this chapter we will prove Theorem 4.0.30. The proof is straightforward, using the method mentioned in the introduction. In Lemma 5.0.40, we give lower bounds for $\varepsilon_2(x)$ for $x \in S_n$, which depend on the number of short cycles in x. This lemma will be used in conjunction with Lemma 5.0.43 to show that for many of the r-tuples E which satisfy the assumptions of Theorem 1.1.2 we have $g_2(E) > cn$ and $3g_2(E) > |E|$. The r-tuples to which this method cannot be applied contain many elements with few fixed points. The indices of such elements are large, as shown in Lemma 5.0.41, and this will be used to again draw the desired conclusions about $g_2(E)$. The proof is very long and tedious, as there are many cases (that is, conditions on E) to consider, including some which give $g_2(E) = 0$.

We will use the following stronger version of Lemma 2.0.19.

LEMMA 5.0.36. *Let $t \in S_n$ be a transposition and let $u \in S_n$ be the product of two commuting transpositions. If $x \in S_n$ is not conjugate to one of $1, t$ then*
$$i_2(x) \geq i_2(u) = 2n - 6.$$

PROOF. Let $y \in S_n$ have shape $1^{n-3}3^1$. If x is not conjugate to one of $1, t, x, y$ then we may assume that the two conditions
- every fixed point of x is also fixed by u, and
- if $(\alpha\beta)$ is a 2-cycle of x then either $(\alpha\beta)$ is a cycle of u or u fixes both α and β

hold. It follows that every 2-set from $[n]$ which is fixed by x is also fixed by u. Since every orbit of u on $\binom{[n]}{2}$ has size at most two, we see that for each $X \in \binom{[n]}{2}$ the $\langle x \rangle$-orbit containing X is larger than the $\langle u \rangle$-orbit containing X. It follows that $o_2(x) \leq o_2(u)$ and $i_2(x) \geq i_2(u)$.

Now Burnside's lemma gives
$$o_2(u) = \frac{1}{2}\left[\binom{n}{2} + \binom{n-4}{2} + 2\right],$$
so
$$i_2(u) = \frac{1}{2}\left[\binom{n}{2} - \binom{n-4}{2} - 2\right] = 2n - 6,$$
and
$$o_2(y) = \frac{1}{3}\left[\binom{n}{2} + 2\binom{n-3}{2}\right],$$
so
$$i_2(y) = \frac{2}{3}\left[\binom{n}{2} - \binom{n-3}{2}\right] = 2n - 4.$$

□

We will also need the elementary results below, the first of which follows immediately from the definitions.

PROPOSITION 5.0.37. *For all $x \in S_n$, we have $f_2(x) = \binom{f_1(x)}{2} + c_2(x)$.*

LEMMA 5.0.38. *Let $x \in S_n$ have shape $(\lambda_1, \ldots, \lambda_r)$. Then*
$$o_2(x) = \sum_{1 \leq i \leq r} \left\lfloor \frac{\lambda_i}{2} \right\rfloor + \sum_{1 \leq i < j \leq r} gcd(\lambda_i, \lambda_j).$$

PROOF. Let $x_1, \ldots, x_r$ be the cycles of x, with $|supp(x_i)| = \lambda_i$. Applying Burnside's lemma to the action of x_i on 2-sets from $supp(x_i)$, we see that the number of orbits of x on these 2-sets is $\lfloor \frac{\lambda_i}{2} \rfloor$. If $i < j$ then $x_i x_j$ acts regularly on each orbit of 2-sets which consist of one element of $supp(x_i)$ and one element of $supp(x_j)$, so each orbit contains $|x_i x_j| = lcm(\lambda_i, \lambda_j)$ 2-sets. Since there are $\lambda_i \lambda_j$ such 2-sets, there are $gcd(\lambda_i, \lambda_j)$ orbits. □

LEMMA 5.0.39. *Let $(\lambda_1, \ldots, \lambda_r) \in P(n)$. Then*
$$\sum_{1 \leq i < j \leq r} gcd(\lambda_i, \lambda_j) \leq \frac{(r-1)n}{2}.$$

PROOF. The claim of the lemma holds when n is small, and we proceed by induction on n. By inductive hypothesis and the fact that $gcd(\lambda_i, \lambda_r) \leq \lambda_r$, we have
$$\sum_{1 \leq i < j \leq r} gcd(\lambda_i, \lambda_j) \leq \frac{(r-2)(n-\lambda_r)}{2} + (r-1)\lambda_r = \frac{(r-1)n - (n - r\lambda_r)}{2}.$$
The lemma now follows from the fact that $r\lambda_r \leq n$. □

Here is our key lemma.

LEMMA 5.0.40. *Let $x \in S_n$ with $n \geq 5$. Let $d = |x|$.*
(1) *If x has cycle shape $1^a 2^{(n-a)/2}$ then $\varepsilon_2(x) = \frac{(a-1)(n-a)}{2n(n-1)}$. In particular, if $f_1(x) = n - 2$ then $\varepsilon_2(x) = \frac{n-3}{n(n-1)}$.*
(2) *If x has cycle shape $1^b 3^{(n-b)/3}$ then $\varepsilon_2(x) = \frac{2b(n-b)}{3n(n-1)}$. In particular, if $f_1(x) = n - 3$ then $\varepsilon_2(x) = \frac{2n-6}{n(n-1)}$.*
(3) *If $5 \leq f_1(x) \leq n - 4$ then $\varepsilon_2(x) \geq \frac{2n-10}{n(n-1)}$, and if $d \geq 4$ then $\varepsilon_2(x) \geq \frac{3n-15}{n(n-1)}$.*
(4) *If $f_1(x) = 4$ then $\varepsilon_2(x) \geq \frac{3n-12}{2n(n-1)}$, and if $d \geq 4$ then $\varepsilon_2(x) \geq \frac{9n-36}{4n(n-1)}$.*
(5) *If $f_1(x) = 3$ then $\varepsilon_2(x) \geq \frac{d-1}{d} \left[\frac{2n-6}{n(n-1)} \right] \geq \frac{n-3}{n(n-1)}$.*
(6) *If $f_1(x) = 2$ then $\varepsilon_2(x) \geq \frac{d-1}{d} \left[\frac{n-2}{n(n-1)} \right] \geq \frac{n-2}{2n(n-1)}$.*
(7) *Say $f_1(x) = 1$. Let s, t be the lengths of the two shortest nontrivial cycles in x (breaking ties arbitrarily), with $s \leq t$.*
  (a) *If x has shape $1^1 s^a$ then*
$$\varepsilon_2(x) = \begin{cases} \frac{s-2}{sn} & , s \text{ even} \\ \frac{s-1}{sn} & , s \text{ odd}. \end{cases}$$

(b) If x has shape $1^1 s^a (2s)^b$ with $a, b > 0$ then
$$\varepsilon_2(x) = \begin{cases} \frac{(a+2)s-2}{2sn} - \frac{a(as+2)}{2n(n-1)} & , s \text{ even} \\ \frac{(a+2)s-2}{2sn} - \frac{a^2 s}{2n(n-1)} & , s \text{ odd.} \end{cases}$$

(c) If s is odd and x has shape $1^1 s^a (3s)^b$ then
$$\varepsilon_2(x) = \frac{1}{n} + \frac{2a(n - as - 2)}{3n(n-1)} - \frac{1}{3sn}.$$

(d) If s is even and x has shape $1^1 s^a (\frac{3s}{2})^b (3s)^c$ with $a > 0$ and $b + c > 0$ then
$$\varepsilon_2(x) = \begin{cases} \frac{1}{n} + \frac{2a}{3n} - \frac{2a(as+2)}{3n(n-1)} + \frac{b}{2n} - \frac{b(3bs+4)}{4n(n-1)} - \frac{2}{3sn} & , 4 | s \\ \frac{1}{n} + \frac{2a}{3n} - \frac{2a(as+2)}{3n(n-1)} + \frac{b}{2n} - \frac{3b^2 s}{4n(n-1)} - \frac{2}{3sn} & , 4 \nmid s. \end{cases}$$

(e) For any d, we have $\varepsilon_2(x) \geq (\frac{1}{s} - \frac{1}{d}) \frac{s(n-1-s)}{n(n-1)}$. In particular, if $d \geq 4s$ then $\varepsilon_2(x) \geq \frac{3(n-1-s)}{4n(n-1)}$.

(f) If $(s, t) = (2, 2)$ then
$$\varepsilon_2(x) \geq \left(\frac{1}{2} - \frac{1}{d}\right) \frac{4n - 20}{n(n-1)}.$$

(g) If $s = 2$ and $c_3(x) > 0$ then
$$\varepsilon_2(x) \geq \frac{2n - 9}{n(n-1)} - \frac{6n - 36}{dn(n-1)}.$$

(h) If $(s, t) = (2, 4)$ then
$$\varepsilon_2(x) \geq \frac{2n - 12}{n(n-1)} - \frac{6n - 42}{dn(n-1)}.$$

(i) If $(s, t) = (2, 5)$ then
$$\varepsilon_2(x) \geq \frac{2n - 11}{n(n-1)} - \frac{10n - 80}{dn(n-1)}.$$

(j) If $(s, t) = (3, 3)$ then
$$\varepsilon_2(x) \geq \left(\frac{1}{3} - \frac{1}{d}\right) \frac{6n - 42}{n(n-1)}.$$

(8) Say $f_1(x) = 0$. Let s, t be the lengths of the two shortest cycles in x (breaking ties arbitrarily), with $s \leq t$.

(a) If x has shape s^a then
$$\varepsilon_2(x) = \begin{cases} -\frac{1}{s(n-1)} & , s \text{ even} \\ 0 & , s \text{ odd.} \end{cases}$$

(b) If x has shape $s^a (2s)^b$ with $a, b > 0$ then
$$\varepsilon_2(x) = \begin{cases} \frac{(as-1)(n-as)-2as}{2sn(n-1)} & , s \text{ even} \\ \frac{(as-1)(n-as)}{2sn(n-1)} & , s \text{ odd.} \end{cases}$$

(c) If s is odd and x has shape $s^a (3s)^b$ with $a, b > 0$ then
$$\varepsilon_2(x) = \frac{2a(n - as)}{3n(n-1)}.$$

(d) *If s is even and x has shape $s^a(\frac{3s}{2})^b(3s)^c$ with $a > 0$ and $b + c > 0$ then*
$$\varepsilon_2(x) = \begin{cases} \frac{2a}{3n} - \frac{2a^2 s}{3n(n-1)} + \frac{b}{2n} - \frac{3b^2 s}{4n(n-1)} - \frac{1}{3s(n-1)} & , 4 \mid s \\ \frac{2a}{3n} - \frac{2a^2 s}{3n(n-1)} + \frac{b}{2n} - \frac{b(3bs-4)}{4n(n-1)} - \frac{1}{3s(n-1)} & , 4 \nmid s. \end{cases}$$

(e) *For any d, we have*
$$\varepsilon_2(x) \geq \left(\frac{1}{s} - \frac{1}{d}\right) \frac{(s-1)(n-s)}{n(n-1)} - \frac{1}{n(n-1)} \sum_{2 \mid m} c_m(x).$$

(f) *If $(s,t) = (2,2)$ then*
$$\varepsilon_2(x) \geq \left(\frac{1}{2} - \frac{1}{d}\right) \frac{3n - 12}{n(n-1)} - \frac{1}{n(n-1)} \sum_{2 \mid m} c_m(x).$$

(g) *If $(s,t) = (2,2)$ and $c_3(x) > 0$ then*
$$\varepsilon_2(x) \geq \frac{4n - 15}{3n(n-1)} + \left(\frac{1}{6} - \frac{1}{d}\right) \frac{4n - 20}{n(n-1)} - \frac{1}{n(n-1)} \sum_{m \in J_3} c_m(x),$$

where for any odd integer l we define
$$J_l := \{m \in \mathbb{N} : m \equiv 2 \bmod 4, l \nmid m\}.$$

(h) *If $(s,t) = (2,3)$ then*
$$\varepsilon_2(x) \geq \frac{2n - 5}{3n(n-1)} + \left(\frac{1}{6} - \frac{1}{d}\right) \frac{4n - 20}{n(n-1)} - \frac{1}{n(n-1)} \sum_{m \in J_3} c_m(x).$$

Moreover, if n is even or $c_3(x) \geq 2$ then
$$\varepsilon_2(x) \geq \frac{17n - 56}{30n(n-1)} + \left(\frac{1}{6} - \frac{1}{d}\right) \frac{4n - 20}{n(n-1)},$$

while if n is odd and $c_3(x) = 1$ then
$$\varepsilon_2(x) \geq \frac{17n - 65}{30n(n-1)} + \left(\frac{1}{6} - \frac{1}{d}\right) \frac{4n - 20}{n(n-1)}.$$

(i) *If $(s,t) = (2,4)$ then*
$$\varepsilon_2(x) \geq \frac{3n - 16}{2n(n-1)} - \frac{5n - 30}{dn(n-1)} - \frac{1}{n(n-1)} \sum_{2 \mid m} c_m(x).$$

(j) *If $(s,t) = (2,5)$ then*
$$\varepsilon_2(x) \geq \frac{7n - 35}{5n(n-1)} - \frac{8n - 56}{dn(n-1)} - \frac{1}{n(n-1)} \sum_{m \in J_5} c_m(x).$$

(k) *If $(s,t) = (3,3)$ then*
$$\varepsilon_2(x) \geq \left(\frac{1}{3} - \frac{1}{d}\right) \frac{5n - 30}{n(n-1)} - \frac{1}{n(n-1)} \sum_{m \in I_3} c_m(x)$$
$$\geq \left(\frac{1}{3} - \frac{1}{d}\right) \frac{5n - 30}{n(n-1)} - \frac{n - 6}{4n(n-1)},$$

where for any odd integer l, we define
$$I_l := \{m \in \mathbb{N} : 2 \mid m, l \nmid m\}.$$

(l) If $(s,t) = (3,4)$ then
$$\varepsilon_2(x) \geq \frac{2n-7}{2n(n-1)} + \left(\frac{1}{12} - \frac{1}{d}\right)\frac{6n-42}{n(n-1)} - \frac{1}{n(n-1)}\sum_{m \in I_3^8} c_m(x),$$
where
$$I_l^8 := \{m \in I_l : 8 \nmid m\}.$$

(m) If $(s,t) = (3,5)$ then
$$\varepsilon_2(x) \geq \frac{16n-64}{15n(n-1)} + \left(\frac{1}{15} - \frac{1}{d}\right)\frac{7n-56}{n(n-1)} - \frac{1}{n(n-1)}\sum_{m \in I_3 \cap I_5} c_m(x)$$
$$\geq \frac{16n-64}{15n(n-1)} + \left(\frac{1}{15} - \frac{1}{d}\right)\frac{7n-56}{n(n-1)} - \frac{n-8}{8n(n-1)}.$$

(n) If $(s,t) = (4,4)$ then
$$\varepsilon_2(x) \geq \left(\frac{1}{4} - \frac{1}{d}\right)\frac{7n-56}{n(n-1)} - \frac{1}{n(n-1)}\sum_{2 \mid m} c_m(x).$$

(o) If $(s,t) = (4,5)$ then
$$\varepsilon_2(x) \geq \frac{6n-27}{5n(n-1)} + \left(\frac{1}{20} - \frac{1}{d}\right)\frac{8n-72}{n(n-1)} - \frac{1}{n(n-1)}\sum_{m \in I_5^8} c_m(x).$$

(p) If $(s,t) = (5,5)$ then
$$\varepsilon_2(x) \geq \left(\frac{1}{5} - \frac{1}{d}\right)\frac{9n-90}{n(n-1)} - \frac{1}{n(n-1)}\sum_{m \in I_5} c_m(x).$$

(9) If $\varepsilon_2(x) < 0$ then x has shape s^a with s even. If $\varepsilon_2(x) = 0$ then the shape of x is one of s^a with s odd, $1^1 2^{(n-1)/2}$ or $2^1 4^1$.

PROOF. Claims 1,2,7a-7d and 8a-8d follow from direct computation of $o_1(x)$ and $o_2(x)$ for all x described in those claims. Now fix $n \geq 5$ and define $\chi : \mathbf{R} \to \mathbf{R}$ by
$$\chi(f) := \frac{f}{n} - \frac{n + f(f-2)}{n(n-1)} = \frac{(f-1)(n-f)}{n(n-1)}.$$
For $x \in S_n$ we have $c_2(x) \leq \frac{n - f_1(x)}{2}$, and it follows from Proposition 5.0.37 that
$$\delta_{1,2}(x) = \frac{f_1(x)}{n} - \frac{\binom{f_2(x)}{2}}{\binom{n}{2}} \geq \chi(f_1(x)).$$
Note that χ is a quadratic polynomial in f with negative leading coefficient. Therefore, on any closed interval $[a,b]$, χ is minimized at one of the endpoints.

Assume $f_1(x) \in \{2,3\}$. Then $f_1(x) \leq f_1(x^j) \leq n-2$ for $1 \leq j < d$. Since $\chi(f_1(x)) \leq \chi(n-2)$, it follows from Proposition 2.0.9 that
$$\varepsilon_2(x) \geq \frac{d-1}{d}\chi(f_1(x)).$$
Now the condition $n \geq 5$ guarantees $\chi(f_1(x)) \geq 0$, so the right side of the above inequality is minimized when $d = 2$. Claims 5 and 6 follow.

Now assume that $f_1(x) = 4$. If no power of x has $n - 2$ fixed points then we can argue as we did when proving claims 5 and 6 to prove claim 4 in this case. If $f_1(x^{d/2}) = n - 2$ then $c_2(x) = 1$ and every cycle of x other than the unique transposition has odd length. If $n = 6$ then $f_1(x) = n - 2$ and direct calculation shows that $\varepsilon_2(x)$ is equal to the lower bound given in claim 4. If $n > 6$ then $d \geq 6$, and Proposition 2.0.9 gives

$$\varepsilon_2(x) \geq \frac{1}{d} \left[(d-2)\chi(4) + \chi(n-2) \right].$$

Under the given conditions, the right side of the above inequality is minimized when $d = 6$, giving

$$\varepsilon_2(x) \geq \frac{7n - 27}{3n(n-1)} > \frac{9n - 36}{4n(n-1)},$$

and claim 4 follows.

Now assume $5 \leq f_1(x) \leq n - 4$. If $5 \leq f_1(x^j) \leq n - 4$ for $1 \leq j < d$ then we can argue as we did when proving claims 5 and 6 to prove claim 3. Otherwise, we have $f_1(x^{d/2}) = n - 2$ or $f_1(x^{d/3}) = f_1(x^{2d/3}) = n - 3$, or both of the conditions just listed hold. If only the first condition holds then $d \geq 6$ and arguing as we did in the case $f_1(x) = 4$ we get

$$\varepsilon_2(x) \geq \frac{1}{6} \left[4\chi(5) + \chi(n-2) \right] = \frac{9n - 43}{3n(n-1)}.$$

If only the second condition holds then every cycle of x other than the unique 3-cycle has length not divisible by three. Since $f_1(x) \neq n - 3$, we have $d \geq 6$ and

$$\varepsilon_2(x) \geq \frac{1}{6} \left[3\chi(5) + 2\chi(n-3) \right] = \frac{3n - 14}{n(n-1)}.$$

If both conditions hold then every cycle of x other than the unique transposition and the unique 3-cycle has length prime to six. Therefore, either x has shape $1^{n-5}2^1 3^1$ or $d \geq 30$. In the first case, direct calculation gives

$$\varepsilon_2(x) = \frac{3n - 13}{n(n-1)}$$

and in the second case we get

$$\varepsilon_2(x) \geq \frac{1}{30} \left[26\chi(5) + 2\chi(n-3) + \chi(n-2) \right] = \frac{56n - 275}{15n(n-1)} > \frac{3n - 15}{n(n-1)},$$

the last inequality holding since $n > 5$. Thus claim 3 holds under any combination of the given conditions.

Now assume $f_1(x) = 1$ and let s be as in claim 7. Since every power of x is a product of powers the cycles of x, no nontrivial power of x has more than $n - s$ fixed points. It follows that for each $j \in [d-1]$ such that $s | j$ we have $s + 1 \leq f_1(x^j) \leq n - s$, and

$$\chi(f_1(x^j)) \geq \frac{s(n-1-s)}{n(n-1)}.$$

Since every power of x is fixes some point, we have $\chi(f_1(x^j)) \geq 0$ for all $j \in [d]$. Therefore

$$\varepsilon_2(x) \geq \frac{1}{d} \left(\frac{d}{s} - 1 \right) \frac{s(n-1-s)}{n(n-1)} = \left(\frac{1}{s} - \frac{1}{d} \right) \frac{s(n-1-s)}{n(n-1)}.$$

If s is fixed then this difference increases with d, and claim 7e follows. Say $c_2(x) \geq 2$. If no nontrivial power of x has at least $n - 3$ fixed points then

$$\varepsilon_2(x) \geq \frac{1}{d} \left[\left(\frac{d}{2} - 1 \right) \chi(5) + \frac{d}{2} \chi(1) \right] = \left(\frac{1}{2} - \frac{1}{d} \right) \frac{4n - 20}{n(n-1)}.$$

Since $c_2(x) > 1$, no power of x has exactly $n-2$ fixed points. If $f_1(x^{\frac{d}{3}}) = f_1(x^{\frac{2d}{3}}) = n-3$ then $c_3(x) = 1$ and every cycle in x other than the unique 3-cycle has length prime to three. It follows that

$$\varepsilon_2(x) \geq \frac{1}{d}\left[\left(\frac{d}{2} - 3\right)\chi(5) + 2\chi(n-3) + \frac{d}{6}\chi(4)\right].$$

Now we have $d \geq 6$, so

$$\frac{d}{6}\chi(4) + 2\chi(n-3) \geq \frac{9n-36}{n(n-1)} \geq \frac{8n-40}{n(n-1)} = 2\chi(5),$$

and claim 7f follows. If $c_2(x) \geq 1$ and $c_3(x) \geq 1$ then

$$f_1(x^j) \geq \begin{cases} 6 & j \equiv 0 \bmod 6, \\ 4 & j \equiv 3 \bmod 6, \\ 3 & j \equiv 2, 4 \bmod 6, \\ 1 & j \equiv 1, 5 \bmod 6. \end{cases}$$

Accounting simultaneously for the possibilities $f_1(x^{\frac{d}{2}}) \in \{n-2, n-4\}$, we get

$$\varepsilon_2(x) \geq \frac{1}{d}\left[\left(\frac{d}{6} - 2\right)\chi(6) + \chi(n-4) + \frac{d}{3}\chi(3) + \left(\frac{d}{6} - 1\right)\chi(4) + \chi(n-2)\right],$$

and claim 7g follows. Claims 7h, 7i and 7j are proved similarly. We handle the case where $(s, t) = (2, 4)$ and some power of x has $n-5$ fixed points by noting that in that case we have

$$4\chi(n-5) + \frac{d}{10}\chi(6) + \frac{d}{20}\chi(n-4) \geq 4\chi(7) + \frac{d}{20}\chi(3).$$

Now assume $f_1(x) = 0$ and let s be as in claim 8. Then no nontrivial power of x has more than $n - s$ fixed points, so for each $j \in [d-1]$ such that $s | j$ we have $s \leq f_1(x^j) \leq n - s$, and

$$\chi(f_1(x^j)) \geq \frac{(s-1)(n-s)}{n(n-1)}.$$

Let J be the set of elements of $[d]$ which are not divisible by s. To prove claim 8e, we note that

$$\sum_{s \in J} \frac{f_1(x^j)}{n} - \frac{f_2(x^j)}{\binom{n}{2}} \geq -\sum_{f_1(x^j)=0} \frac{c_2(x^j)}{\binom{n}{2}} \geq -\sum_{j=1}^{d} \frac{c_2(x^j)}{\binom{n}{2}},$$

so

$$\varepsilon_2(x) \geq \frac{1}{d}\left[\left(\frac{d}{s} - 1\right)\frac{(s-1)(n-s)}{n(n-1)} - \frac{1}{\binom{n}{2}}\sum_{j=1}^{d} c_2(x^j)\right].$$

For each cycle of x having even length m and each $i \in \left[\frac{d}{m}\right]$, there are $\frac{m}{2}$ 2-cycles which appear in $x^{im/2}$. Any 2-cycle which appears in any power of x arises in this way, and

$$\sum_{j=1}^{d} c_2(x^j) = \sum_{2|m} c_m(x) \frac{d}{m} \frac{m}{2} = \frac{d}{2} \sum_{2|m} c_m(x).$$

Claim 8e follows. Claim 8f is proved similarly, noting that if $c_2(x) \geq 2$ then each nontrivial even power y of x satisfies $4 \leq f_1(y) \leq n - 3$.

If $c_2(x) \geq 2$ and $c_3(x) \geq 1$ then we have

- $5 \leq f_1(x^j) \leq n-4$ whenever $j \equiv 0 \bmod 6$,
- $4 \leq f_1(x^j) \leq n-3$ whenever $j \equiv 2, 4 \bmod 6$, and
- $3 \leq f_1(x^j) \leq n-4$ whenever $j \equiv 3 \bmod 6$.

This gives
$$\varepsilon_2(x) \geq \frac{1}{d}\left[\left(\frac{d}{6}-1\right)\chi(5) + \frac{d}{3}\chi(4) + \frac{d}{6}\chi(3) - \sum_{j \equiv 1,5 \bmod 6} \frac{c_2(x^j)}{\binom{n}{2}}\right]$$

Arguing as we did in the proof of claim 8e, we see that
$$\sum_{j \equiv 1,5 \bmod 6} c_2(x^j) = \frac{d}{2}\sum_{m \in J_3} c_m(x),$$
and claim 8g follows. Claim 8h is proved in a similar manner, noting that if $c_2(x) = 1$ and $c_3(x) \geq 2$ then
$$\sum_{m \in J_3} c_m(x) \leq 1 + \frac{n-8}{10}.$$
Similar arguments prove claims 8i through 8p. Care must be (and was) taken to account for the fact that nontrivial powers of x may have many fixed points.

Finally, we show that claim 9 follows from the other claims. Using direct calculation, we see that claim 9 holds whenever x satisfies the conditions of any of claims 1-6, 7a-e and 8a-d. Therefore, if x is a counterexample to claim 9 then $f_1(x) = 0$ and $d \geq 4s$, where s is as in claim 8. Under these conditions, x has at most $\frac{n}{s}$ cycles on $[n]$, and it follows from claim 8e that
$$\varepsilon_2(x) \geq \frac{(3s-7)n - 3s(s-1)}{4sn(n-1)}.$$
Fix n and set $\phi(s) = (3s-7)n - 3s(s-1)$. Since ϕ is quadratic in s with negative leading coefficient, ϕ is minimized at the endpoints of any closed interval on which it is defined. We have $2 \leq s < \frac{n}{2}$. Direct calculation shows that $\phi(3) > 0$ whenever $n > 9$, $\phi(4) > 0$ whenever $n > 7$ and $\phi(\frac{n}{2}) > 0$ whenever $n > 7$. If $s = 3$ and $d \geq 4s$ and $7 \leq n \leq 9$ then x has shape 3^14^1 or 3^15^1 and $\varepsilon_2(x) > 0$.

It remains to examine the case where $s = 2$ and $d \geq 8$. First assume that $c_3(x) = 0$. Let $a = c_2(x)$, so $1 \leq a \leq \frac{n-5}{2}$, and let $b = \sum_{j>3} c_j(x)$, so $b \leq \frac{n-2a}{4}$. By Lemmas 5.0.38 and 5.0.39, we have
$$\begin{aligned}\varepsilon_2(x) &\geq \frac{a+b}{n} - \frac{1}{\binom{n}{2}}\left(\frac{n}{2} + 2\left[\binom{a}{2} + ab\right] + \frac{(b-1)(n-2a)}{2}\right) \\ &= \frac{a(n-2a-1) - b(2a+1)}{n(n-1)}.\end{aligned}$$
Since $b \leq \frac{n-2a}{4}$ we have
$$\varepsilon_2(x) \geq \frac{(2a-1)n - 2a(2a+1)}{4n(n-1)} = \frac{(n-2a)(2a-1) - 4a}{4n(n-1)}.$$
Since $1 \leq a \leq \frac{n-5}{2}$, we have
$$\varepsilon_2(x) \geq \frac{6a-5}{4n(n-1)} > 0.$$
If $c_3(x) > 0$ then it follows from claims 8g and 8h that $\varepsilon_2(x) > 0$. $\square$

5. ACTIONS ON 2-SETS

LEMMA 5.0.41. *Let $x \in S_n$ with $n \geq 5$.*
0. *If $f_1(x) = 0$ then $i_2(x) \geq \frac{1}{2}\binom{n}{2} - \frac{n}{4}$.*
1. *If $f_1(x) = 1$ then $i_2(x) \geq \frac{1}{2}\binom{n}{2} - \frac{n-1}{4}$.*
2. *If $f_1(x) = 2$ then $i_2(x) \geq \frac{1}{2}\binom{n}{2} - \frac{n}{4}$.*
3. *If $f_1(x) = 3$ then $i_2(x) \geq \frac{1}{2}\binom{n}{2} - \frac{n+3}{4}$.*
4. *If $f_1(x) = 4$ then $i_2(x) \geq \frac{1}{2}\binom{n}{2} - \frac{n+8}{4}$.*

PROOF. The given lower bounds on $i_2(x)$ are obtained by first noting that $o_1(x) \leq \frac{n+f_1(x)}{2}$, so that $i_1(x) \geq \frac{n-f_1(x)}{2}$, and then applying Lemma 5.0.40. □

For the rest of this chapter, we assume that $E = (x_1, \ldots, x_r)$ is an r-tuple of nonidentity elements of S_n satisfying the assumptions of Theorem 4.0.30, that is,

(R) $r \geq 5$,
(N) $n > 20$,
(A) $A_n \leq \langle E \rangle$, and
(P) $\prod_{i=1}^r x_i = 1$.

Define
$$T := \left\{ x \in E : x \text{ has shape } 1^{n-2}2^1 \right\},$$
$$A := \left\{ x \in E : f_1(x) \leq 4 \right\},$$
$$a := |A|$$
and
$$B := \{ x \in A : x \text{ is homocyclic of even order} \}.$$
For $0 \leq i \leq 4$ define
$$A_i := \{ x \in A : f_1(x) = i \}$$
and
$$a_i := |A_i|$$
(so $a = \sum_{i=0}^{4} a_i$).

The next lemma follows immediately from Proposition 2.0.16.

LEMMA 5.0.42. *If $g_1(E) = 0$ then $|B| \leq 1$.*

We will prove Theorem 4.0.30 by examining conditions on $|T|$ and $|A|$ which exhaust all possibilities. In each case, we will see that either condition 2 of the theorem holds or $g_2(E) - g_1(E)$ is bounded below by a polynomial in n and r (or in n, r and $|A|$) which specializes, upon substitution of appropriate values for r and $|A|$, to a polynomial $\Psi(n)$ of degree least one with positive leading coefficient. Moreover, we will in see that $\Psi(n) > 2$ whenever $n > 20$ (usually this will be apparent and not mentioned explicitly). Since there are finitely many cases, this gives the existence of the constant c described in the theorem. A similar strategy will be employed in each case to show that $g_2(E) - \frac{r}{3} > 0$ whenever condition 2 does not hold. We will often employ elementary calculus to obtain our results, and we will not show our calculations when doing so.

The next lemma is an immediate consequence of Lemma 2.0.6

LEMMA 5.0.43. *We have*

$$(5.1) \qquad g_2(E) - g_1(E) = \frac{n-3}{2}(g_1(E) - 1) + \frac{n(n-1)}{4} \sum_{x \in E} \varepsilon_2(x),$$

and
$$(5.2) \quad g_2(E) - \frac{r}{3} = 1 - \frac{r}{3} + \frac{n-1}{2}(g_1(E) - 1) + \frac{n(n-1)}{4}\sum_{x \in E}\varepsilon_2(x).$$

LEMMA 5.0.44. *Define*
$$\Psi(n,r,a) := \frac{1}{8}\left[4(n-3)(r-n) + a(n^2 - 8n + 12)\right]$$
and
$$\Gamma(n,r,a) := \frac{1}{8}\left[r(4n-8) - 4n^2 + 4n + a(n^2 - 6n)\right] + 1 - \frac{r}{3}.$$
Then
$$(5.3) \quad g_2(E) - g_1(E) \geq \Psi(n,r,|A|)$$
and
$$(5.4) \quad g_2(E) - \frac{r}{3} \geq \Gamma(n,r,|A|).$$

PROOF. Direct calculation gives
$$g_1(E) - 1 \geq \frac{r - a - 2n}{2} + \sum_{i=0}^{4} a_i \frac{n-i}{4}.$$
Using Lemma 5.0.40, we get
$$\sum_{x \in E}\varepsilon_2(x) \geq \frac{1}{2n(n-1)}\left[(r-a)(2n-6) + \sum_{i=0}^{4}(i-1)a_i(n-i)\right].$$
Combining the two equalities just given with equation (5.1) of Lemma 5.0.43, we get
$$g_2(E) - g_1(E) \geq \frac{1}{8}\left[4(n-3)(r-n) + \sum_{i=0}^{4}(n^2 - 8n + 12 + 4i - i^2)\right],$$
and inequality (5.3) follows. Inequality (5.4) is proved similarly, using equation (5.2) of Lemma 5.0.43 in place of equation (5.1). □

COROLLARY 5.0.45. *Theorem 4.0.30 is true if we assume that $|A| \geq 5$ or that $|A| = 4$ and $r \geq 6$.*

PROOF. It is easy to see that (given $n > 20$) both Ψ and Γ are increasing functions of a and r. Direct calculation gives
$$\Psi(n,6,4) = \frac{n-6}{2} > 2$$
and
$$\Psi(n,5,5) = \frac{n(n-8)}{2} > \frac{n}{4} > 2,$$

so the desired conclusions on $g_2(E) - g_1(E)$ follow from Lemma 5.0.44. Also (since $g_2(E) \geq g_2(E) - g_1(E)$), we have $g_2(E) - \frac{r}{3} > 0$ when the given conditions hold and $r \leq 6$. Direct calculation also gives

$$\Gamma(n, 7, 4) = n - \frac{25}{3} > 0$$

and we are done. $\square$

LEMMA 5.0.46. *Theorem 4.0.30 is true if we assume that $r = 5$ and $|A| = 4$.*

PROOF. Note first that to prove $g_2(E) - \frac{r}{3} > 0$ it suffices to prove that $g_2(E) - g_1(E) > 1$. Let t be the unique element of $E \setminus A$. Assume first that t is not a transposition. Then $i_1(t) > 1$, so

$$2(g_1(E) + n - 1) \geq 2 + \sum_{i=0}^{4} a_i \frac{n-i}{2},$$

so

(5.5) $$g_1(E) - 1 \geq 1 - \frac{1}{4} \sum_{i=0}^{4} a_i i.$$

Using Lemma 5.0.40(1-3), we see that

$$\varepsilon_2(t) \geq \frac{2n - 10}{n(n-1)}$$

and

(5.6) $$\frac{n(n-1)}{4} \sum_{x \in E} \varepsilon_2(x) \geq \frac{1}{8} \left[4n - 20 + \sum_{i=0}^{4} a_i(i-1)(n-i) \right].$$

Combining inequalities (5.5) and (5.6) with Lemma 5.0.43, we get

$$g_2(E) - g_1(E) \geq \frac{1}{8} \left[4n - 32 + \sum_{i=0}^{4} a_i(4i - i^2) \right] \geq \frac{n-8}{2},$$

and our claim holds in this case.

Now assume that t is a transposition, so $i_1(t) = 1$ and

$$\varepsilon_2(t) = \frac{n-3}{n(n-1)}.$$

Say $g_1(E) \geq 2$. Lemmas 5.0.40(8a,9) and 5.0.43 give

$$g_2(E) - g_1(E) \geq \frac{3n-9}{4} - 4\frac{n}{8} = \frac{n-9}{4} > 2.$$

So, we may assume that $g_1(E) \leq 1$. Let $A = \{x_1, x_2, x_3, x_4\}$ with $f_1(x_i) \geq f_1(x_{i+1})$ for $i = 1, 2, 3$ and set

$$\bar{v} := (f_1(x_1), f_1(x_2), f_1(x_3), f_1(x_4)).$$

Note that each entry of $\bar{v}$ lies in $\{0, 1, 2, 3, 4\}$.

Say $g_1(E) = 1$. Lemma 5.0.43 gives

$$g_2(E) - g_1(E) = \frac{n-3}{4} + \frac{n(n-1)}{4} \sum_{x \in A} \varepsilon_2(x).$$

If n is odd then $\varepsilon_2(x) \geq 0$ for all $x \in A$ by Lemma 5.0.40(8), so
$$g_2(E) - g_1(E) \geq \frac{n-3}{4} > 2.$$
So, assume n is even. Note first that
$$2n = 2(g_1(E) + n - 1) = \sum_{x \in E} n - o_1(x) = 1 + 4n - \sum_{x \in A} o_1(x),$$
so
$$\sum_{x \in A} o_1(x) = 2n + 1.$$
Now for any $x \in A$ we have
$$o_1(x) \leq \lfloor \frac{n + f_1(x)}{2} \rfloor,$$
with equality if and only if either $f_1(x)$ is even and $|x| = 2$ or $f_1(x)$ is odd and x has shape $1^i 2^{(n-3-i)/2} 3^1$. It follows that one of the cases in the table below must occur.

$\overline{v}$	lower bound on $g_2(E) - g_1(E)$	Lemmas used
$(4, \cdot, \cdot, \cdot)$	$(n-9)/4$	$5.0.40(4, 8a, 9)$
$(3, \cdot, \cdot, \cdot)$	$(2n-15)/8$	$5.0.40(2, 5, 8a, 9)$
$(2, 2, \cdot, \cdot)$	$(n-5)/4$	$5.0.40(6, 8a, 9)$
$(2, 1, 1, \cdot)$	$(n-4)/4$	$5.0.40(6, 8a, 9)$
$(2, 1, 0, 0)$	$(n-4)/4$	$5.0.40(6, 7d, 8a, 9)$

Note that the lower bound when $\overline{v} = (2,1,0,0)$ is obtained by observing that if $|B| < 2$ then the bound in the case $\overline{v} = (2,1,1,\cdot)$ applies, while if $|B| = 2$ then x_2 has shape $1^1 2^{(n-4)/2} 3^1$ and $g_2(E) - g_1(E) \geq (n-4)/2$. In any case, the claim of our lemma follows.

Finally, assume $g_1(E) = 0$. By Lemma 5.0.42, we have $|B| \leq 1$. By Lemmas 5.0.43 and 5.0.40(1), we have
$$g_2(E) - g_1(E) = -\frac{n-3}{4} + \frac{n(n-1)}{4} \sum_{x \in A} \varepsilon_2(x).$$
Also,
$$2(n-1) = 2(g_1(E) + n - 1) = \sum_{x \in E} n - o_1(x) = 1 + 4n - \sum_{x \in A} o_1(x),$$
so
$$\sum_{x \in A} o_1(x) = 2n + 3.$$
Using the bounds on $o_1(x)$ described in the case $g_1(E) = 1$, we see that if n is odd then one of the cases in the table below must occur,

$\overline{v}$	lower bound on $g_2(E) - g_1(E)$	Lemmas used
$(4, \cdot, \cdot, \cdot)$	$(5n-24)/16$	$5.0.40(2, 4, 9)$
$(3, 3, \cdot, \cdot)$	$(n-3)/4$	$5.0.40(5, 9)$
$(3, 2, \cdot, \cdot)$	$(n-2)/6$	$5.0.40(5, 6, 9)$
$(3, 1, 1, 1)$	0	$5.0.40(1)$

while if n is even then one of the cases in the next table must occur.

$\overline{v}$	lower bound on $g_2(E) - g_1(E)$	Lemmas used
$(4, 4, \cdot, \cdot)$	$(7n - 18)/8$	5.0.40(4, 8a, 9)
$(4, 3, \cdot, \cdot)$	$(4n - 21)/12$	5.0.40(4, 5, 8a, 9)
$(4, 2, 2, \cdot)$	$(n - 5)/4$	5.0.40(4, 6, 8a, 9)
$(4, 2, 1, \cdot)$	$(n - 4)/2$	5.0.40(4, 6, 7d, 8a, 9)
$(3, \geq 2, \geq 2, \cdot)$	$(5n - 18)/24$	5.0.40(5, 6, 8a, 9)
$(2, 2, 2, > 0)$	$n/8$	5.0.40(6, 9)
$(2, 2, 2, 0)$	0	5.0.40(1)

(Note that if n is even and $\overline{v} = (4, 2, 1, \cdot)$ then x_3 has shape $1^1 2^{(n-4)/2} 3^1$ and that if n is even and $\overline{v} = (4, 2, 0, 0)$ then $|B| = 2$, which is impossible.)

Now if n is odd and $\overline{v} = (3, 1, 1, 1)$ or if n is even and $\overline{v} = (2, 2, 2, 0)$ then each element of A is an involution and $g_2(E) = g_1(E) = 0$. In fact, in these cases, E is as described in condition (2) of Theorem 4.0.30. Moreover, by Proposition 3.0.24, there exist 5-tuples E of the types under consideration which satisfy conditions (R),(N),(A),(P). In all other cases, the claim of our lemma holds. $\square$

LEMMA 5.0.47. *Theorem 4.0.30 is true if we assume that $|A| = 3$ and $T = \emptyset$.*

PROOF. First assume $g_1(E) > 0$. By equation (5.1) of Lemma 5.0.43 and Lemma 5.0.40(1,2,3,8a,9) we have

$$g_2(E) - g_1(E) \geq \frac{n(n-1)}{4} \sum_{x \in E} \varepsilon_2(x) \geq -\frac{3n}{8} + \frac{(r-3)(n-5)}{2} \geq \frac{5n - 40}{8},$$

and it follows that $g_2(E) - g_1(E) > 2$. A similar argument using equation (5.2) in place of equation (5.1) shows that

$$g_2(E) - \frac{r}{3} \geq 1 + \frac{(r-3)(n-5)}{2} - \frac{3n}{8} - \frac{r}{3},$$

and it follows that $g_2(E) - \frac{r}{3} > 0$.

Now assume $g_1(E) = 0$. By Lemma 5.0.42 we have $|B| \leq 1$. Now equation (5.1) and Lemma 5.0.40(1,2,3,8a,9) give

$$(5.7) \qquad g_2(E) - g_1(E) \geq -\frac{n-3}{2} - \frac{n}{8} + \frac{(r-3)(n-5)}{2} \geq \frac{3n - 28}{8}.$$

It follows that $g_2(E) - g_1(E) > 2$.

Our results about $g_2(E) - g_1(E)$ imply that $g_2(E) - \frac{r}{3} > 0$ when $r \leq 6$, so it remains to show that $g_2(E) - \frac{r}{3} > 0$ whenever $r \geq 7$. Now equation (5.1) of Lemma 5.0.43 and Lemma 5.0.40(1,2,3,8a,9) give

$$g_2(E) - \frac{r}{3} \geq -\frac{n-3}{2} - \frac{n}{8} + \frac{(r-3)(n-5)}{2} - \frac{r}{3}$$

and straightforward calculations give $g_2(E) - \frac{r}{3} > 0$ as desired. $\square$

LEMMA 5.0.48. *Theorem 4.0.30 is true if we assume that $|A| = 3$ and $|T| = 1$.*

PROOF. First assume $g_1(E) > 0$. By equation (5.1) of Lemma 5.0.43 and Lemma 5.0.40(1,2,3,8a,9), we have
$$g_2(E) - g_1(E) \geq -\frac{3n}{8} + \frac{n-3}{4} + \frac{(r-4)(n-5)}{2} \geq \frac{3n-26}{8}.$$
It follows that $g_2(E) - g_1(E) > 2$.

The results on $g_2(E) - g_1(E)$ give $g_2(E) - \frac{r}{3} > 0$ when $r \leq 6$, so assume $r \geq 7$. We have by equation (5.2) of Lemma 5.0.43 and Lemma 5.0.40(1,2,3,8a,9) that
$$g_2(E) - \frac{r}{3} \geq 1 - \frac{3n}{8} + \frac{n-3}{4} + \frac{(r-4)(n-5)}{2} - \frac{r}{3},$$
and it follows that $g_2(E) - \frac{r}{3} > 0$.

Now assume that $g_1(E) = 0$. By Lemma 5.0.42 we have $|B| \leq 1$. Equation (5.2) and Lemma 5.0.40(1,2,3,8a,9) give
$$g_2(E) - g_1(E) > g_2(E) - \frac{r}{3} \geq -\frac{n-3}{2} - \frac{n}{8} + \frac{n-3}{4} + \frac{(r-4)(n-5)}{2} - \frac{r}{3},$$
and it follows that if $r \geq 6$ then $g_2(E) - g_1(E) > g_2(E) - \frac{r}{3} > 2$.

We may (and do) now assume that $r = 5$. To show that $g_2(E) - \frac{r}{3} > 0$ it suffices to show that $g_2(E) > 1$.

Let us review the situation. The set E contains one transposition T, and $E \setminus (T \cup A)$ contains one element w. If $B = \emptyset$ then Equation (5.1) and Lemma 5.0.40(1,2,3,8a,9) give
$$g_2(E) \geq -\frac{n-3}{2} + \frac{n-3}{4} + \frac{n-5}{2} = \frac{n-7}{4},$$
so $g_2(E) > 2$.

Assume now that $B \neq \emptyset$, so B consists of a single element u of even order l. Let $A \setminus B = \{x, y\}$. Equation (5.1) and Lemma 5.0.40(1,8a) now give
$$(5.8) \qquad g_2(E) = -\frac{n-3}{2} + \frac{n-3}{4} - \frac{n}{4l} + \frac{n(n-1)}{4}(\varepsilon_2(w) + \varepsilon_2(x) + \varepsilon_2(y)).$$
If $|w| \geq 4$ then (5.8) and Lemma 5.0.40(5,9) give
$$g_2(E) \geq -\frac{3n-6}{8} + \frac{3n-15}{4} = \frac{3n-24}{8} > 2.$$
Say $|w| = 3$, and let $b = f_1(w) \geq 5$. If $5 \leq b \leq n-6$ then (5.8) and Lemma 5.0.40(2,9) give
$$g_2(E) \geq -\frac{3n-6}{8} + \frac{5n-25}{6} = \frac{11n-82}{24} > 2.$$
If $b = n-3$ then we have
$$g_2(E) \geq -\frac{3n-6}{8} + \frac{n-3}{2} + \frac{n(n-1)}{4}(\varepsilon_2(x) + \varepsilon_2(y)).$$
Lemma 5.0.40(9) gives $g_2(E) > 2$ if $n > 22$. Moreover, if $l > 2$ then we have
$$g_2(E) \geq -\frac{n-3}{4} - \frac{n}{16} + \frac{n-3}{2} = \frac{3n-12}{16} > 2.$$
We must examine the case where $n = 22$ and $l = 2$. In this case, the Riemann-Hurwitz formula (applied to $g_1(E)$) gives
$$o_1(x) + o_1(y) = \frac{n+10}{2} = 16,$$

so we may assume that $o_1(x) \geq 8$. It follows that if $f_1(x) = 0$ then $f_2(x) > 1$ (and $|x| \geq 6$ by Proposition 2.0.16) and that if $f_1(x) = 1$ then either $f_2(x) > 0$ or x has shape $1^1 3^7$. Now Lemma 5.0.40(6,7a,7e,8f) gives

$$\varepsilon_2(x) \geq \frac{n-2}{2n(n-1)}$$

and we have

$$g_2(E) \geq -\frac{3n-6}{8} + \frac{n-3}{2} + \frac{n-2}{8} = \frac{n-4}{4} > 2.$$

Say $|w| = 2$ and let $a = f_1(w)$. Since n is even and $w \notin A \cup T$ we have $6 \leq a \leq n-4$. Assume first that $a \leq n-6$. Then Lemma 5.0.40(1) gives

$$\varepsilon_2(w) \geq \frac{5n-30}{2n(n-1)}.$$

If $l > 2$ then (5.8) and Lemma 5.0.40(9) give

$$g_2(E) \geq -\frac{n-3}{4} - \frac{n}{16} + \frac{5n-30}{8} = \frac{5n-48}{16} > 2.$$

Say $l = 2$. If $a \geq 8$ then Lemma 5.0.40(1) gives

$$\varepsilon_2(w) \geq \frac{3n-21}{n(n-1)},$$

and we have

$$g_2(E) \geq -\frac{3n-6}{8} + \frac{3n-21}{4} = \frac{3n-36}{8} > 2.$$

If $a = 6$ then Lemma 5.0.40(1) and (5.8) give

$$g_2(E) = \frac{n-12}{4} + \frac{n(n-1)}{4}\left(\varepsilon_2(x) + \varepsilon_2(y)\right).$$

Also, the Riemann-Hurwitz formula gives

$$o_1(x) + o_1(y) = n,$$

so we may assume that $o_1(x) \geq \frac{n}{2}$. Since $x \notin B$ we have $f_1(x) \geq 2$ or x has shape $1^1 2^{(n-4)/2} 3^1$. Now Lemma 5.0.40(4,5,6,7d) gives

$$\varepsilon_2(x) \geq \frac{n-2}{2n(n-1)}$$

and it follows that

$$g_2(E) \geq \frac{3n-26}{8} > 2.$$

We are left with the case $a = n-4$. Lemma 5.0.40(1) and (5.8) give

$$g_2(E) = -\frac{n-3}{4} - \frac{n}{4l} + \frac{n-5}{2} + \frac{n(n-1)}{4}\left(\varepsilon_2(x) + \varepsilon_2(y)\right).$$

When $l > 2$ we get

$$g_2(E) \geq \frac{3n-28}{16} > 2.$$

Say $l = 2$. We have

$$g_2(E) = \frac{n-14}{8} + \frac{n(n-1)}{4}\left(\varepsilon_2(x) + \varepsilon_2(y)\right),$$

so $g_2(E) > 2$ when $n > 30$ by Lemma 5.0.40(9). The Riemann-Hurwitz formula gives
$$o_1(x) + o_1(y) = \frac{n+10}{2},$$
so we may assume that $22 \leq n \leq 30$ and $o_1(x) \geq \frac{n+10}{4}$. If $f_1(x) \geq 2$ then Lemma 5.0.40(4,5,6) gives
$$\varepsilon_2(x) \geq \frac{n-2}{2n(n-1)}$$
and Lemma 5.0.40(9) gives
$$g_2(E) \geq \frac{n-8}{4} > 2.$$
If $f_1(x) = 1$ then the conditions on n and $o_1(x)$ show that x has shape $1^1 3^7$ or $1^1 3^9$ or we can apply Lemma 5.0.40(7e) with $s \in \{2,3\}$ and $d \geq 6$. It now follows that
$$\varepsilon_2(x) \geq \frac{n-4}{2n(n-1)}$$
(this bound is obtained by taking $s = 3$ and $d = 6$ in (7e)). Therefore,
$$g_2(E) \geq \frac{n-9}{4} > 2.$$
If $f_1(x) = 0$ then the conditions on n and $o_1(x)$ show that x has shape $2^1 3^8$ or 3^{10} or $f_2(x) \geq 2$. In the last case, $|x| \geq 6$ and x contains at least two odd cycles by Proposition 2.0.16. By Lemma 5.0.40(8d,8f), if x does not have shape 3^{10} then we have
$$\varepsilon_2(x) \geq \frac{n-2}{2n(n-1)},$$
so
$$g_2(E) \geq \frac{n-8}{4} > 2.$$
If x has shape 3^{10} then $\varepsilon_2(x) = 0$ and $o_1(y) = 10$. By Proposition 2.0.16 we know that y does not have shape 3^{10}. It now follows from Lemma 5.0.40(9) that $\varepsilon_2(y) > 0$, and $g_2(E) = 2 + \varepsilon_2(y) > 2$. □

LEMMA 5.0.49. *Theorem 4.0.30 is true if we assume that $|A| = 3$ and $|T| = 2$.*

PROOF. First assume $g_1(E) > 0$. By Lemmas 5.0.43 and 5.0.40(1-6,8a,9) we have
$$g_2(E) - \frac{r}{3} \geq 1 - \frac{r}{3} + \frac{n-3}{2} - \frac{3n}{8} + \frac{(r-5)(n-5)}{2} > 0,$$
and
(5.9) $$g_2(E) - g_1(E) \geq \frac{n-3}{2} g_1(E) - \frac{3n}{8} + \frac{(r-5)(n-5)}{2}.$$
Inequality (5.9) gives $g_2(E) - g_1(E) > 2$ unless $r = 5$, $g_1(E) = 1$ and $n \leq 28$. In this last case, the Riemann-Hurwitz formula gives
$$\sum_{x \in A} o_1(x) = n + 2,$$
and Lemma 5.0.40(8a,9) gives
$$\frac{n(n-1)}{4} \sum_{x \in A} \varepsilon_2(x) \geq -\frac{n+2}{4}.$$

It follows that
$$g_2(E) - g_1(E) \geq \frac{n-8}{4} > 2.$$

Now assume $g_1(E) = 0$. By Lemma 5.0.42 we have $|B| \leq 1$. Equation (5.1) of Lemma 5.0.43 and Lemma 5.0.40(1-6,8a,9) give
$$g_2(E) \geq \frac{(r-5)(n-5)}{2} - \frac{n}{8}.$$

It follows that $g_2(E) - \frac{r}{3} > 0$ (and therefore $g_2(E) > 2$) whenever $r \geq 6$.

So, we assume that $E = A \cup T$. Let $A = \{x, y, z\}$. Equation (5.1) gives

(5.10) $$g_2(E) = \frac{n(n-1)}{4} \left(\varepsilon_2(x) + \varepsilon_2(y) + \varepsilon_2(z) \right).$$

First assume that $f_1(x) \geq 2$. If $B = \emptyset$ then Lemma 5.0.40(4-6,9) gives
$$g_2(E) \geq \frac{n-2}{8} > 2.$$

So, we assume that $z \in B$. Let $d = |x|$ and let $l = |z|$. If it is not the case that $f_1(x) = l = 2$ and $d \neq 3$ then one of the cases in the table below holds.

Conditions on A	Lower bound on $g_2(E)$	Lemmas used	Comments
$f_1(x) = 4$	$(n-6)/4$	5.0.40(4, 8a, 9)	
$f_1(x) = 3$	$(5n-24)/24$	5.0.40(5, 8a, 9)	$2 \mid n \Rightarrow d \geq 3$
$f_1(x) = 2; d, l \geq 4$	$(n-3)/8$	5.0.40(6, 8a, 9)	
$f_1(x) = 2, d = 3$	$(5n-16)/24$	5.0.40(2, 8a, 9)	
$f_1(x) = d = 2, l \geq 4$	$(n-4)/16$	5.0.40(1, 8a, 9)	
$f_1(x) = d = 2, l \geq 4$, $f_1(y) \geq 2$	$(3n-8)/16$	5.0.40(1, 4, 5, 6, 8a, 9)	
$f_1(x) = d = 2, l \geq 6$	$(n-3)/12$	5.0.40(1, 8a, 9)	
$f_1(x) = d = 2, l \geq 8$	$(3n-8)/32$	5.0.40(1, 8a, 9)	
$f_1(x) = d = 2, l \geq 12$	$(5n-12)/48$	5.0.40(1, 8a, 9)	

It now follows that if one of the cases in the table above holds, we have $g_2(E) > 2$ unless $f_1(x) = d = 2$, $f_1(y) \leq 1$ and the pair (n, l) is one of $(24, 4)$, $(24, 6)$, $(24, 8)$, $(28, 4)$, $(32, 4)$ or $(36, 4)$. The Riemann-Hurwitz formula gives
$$o_1(y) = \frac{n}{2} - \frac{n}{l} + 3,$$
and it follows that in each of the cases just listed we have $c_2(y) + c_3(y) > 2$. Moreover, by Proposition 2.0.16, if $f_1(y) = 0$ then either $c_3(y) > 0$ or $|y| \geq 10$. It now follows from Lemma 5.0.40(7a,7f,,7g,8f,8g,8h,8k) that unless $|y| = 3$ and $f_1(y) = 0$, we have
$$\varepsilon_2(y) \geq \frac{7n - 42}{12n(n-1)},$$
this bound arising from taking $d = 6$ in (8k). Now equation (5.10) and Lemma 5.0.40(1,8a) give
$$g_2(E) \geq \frac{n-2}{8} + \frac{7n-42}{48} - \frac{n}{16} > 2.$$

If $f_1(y) = 0$ and $|y| = 3$ then our formula for $o_1(y)$ gives $(n, l) = (36, 4)$. Now applying Corollary 2.0.17 with $x_1 = x$, $x_2 = z$ and $x_3 = y$ gives a contradiction.

We must examine the case where $f_1(x) = l = 2$ and $d \neq 3$. Direct calculation shows that if d is an odd prime then
$$\varepsilon_2(x) = \frac{2(d-1)(n-2)}{dn(n-1)},$$
so if $d > 3$ is prime, we have
$$g_2(E) \geq \frac{8n - 16}{20} - \frac{n}{8} = \frac{11n - 32}{40} > 2.$$
Further direct calculations give the following results.

Shape of x	$\varepsilon_2(x)$	Lower bound on $g_2(E)$
$1^2 2^a 4^b$	$\frac{(2a+5)(n-2a)-10}{4n(n-1)}$	$(3n-10)/16$
$1^2 2^a 3^b 6^c$	$\frac{9n-18+4a(n-2a-5)+3b(n-3b-3)}{6n(n-1)}$	$(n-3)/4$
$1^2 2^a 4^b 8^c$	$\frac{13n-26+6a(n-2a-2b-5)+4b(n-a-4b-5)}{8n(n-1)}$	$(9n-26)/32$

Thus $g_2(E) > 2$ whenever $4 \leq d \leq 8$. If $d \geq 9$ then Lemma 5.0.40(6) gives
$$\varepsilon_2(x) \geq \frac{8n - 16}{9n(n-1)},$$
so
$$g_2(E) \geq \frac{7n - 32}{72} + \frac{n(n-1)}{4} \varepsilon_2(y).$$
This gives $g_2(E) > 2$ when $n > 24$. If $n \in \{22, 24\}$ we cannot have $d = 9$, so $d \geq 10$. It follows that $o_1(x) \leq 10$ when $n = 22$ and $o_1(x) \leq 11$ when $n = 24$. The Riemann-Hurwitz formula now gives $o_1(y) \geq 5$. It follows that if $f_1(y) = 1$ then (in the language of Lemma 5.0.40(8)) $s \leq 5$, while if $f_1(y) = 0$ then $s \leq 4$. In addition, if $f_1(y) = 0$ and $s = 4$ then $d \geq 20$ by Lemma 2.0.16. Now Lemma 5.0.40(4,5,6,7e,8e,8f,8h,8i,8j) gives
$$\varepsilon_2(y) \geq \frac{1}{4(n-1)},$$
(this bound coming from $s = 3, d = 12$ in (8e)), so
$$g_2(E) \geq \frac{9n - 18}{40} - \frac{n}{8} + \frac{n}{16} = \frac{13n - 36}{80} > 2.$$
Now say $l = d = f_1(x) = 2$. Direct calculation gives
$$i_2(x) = i_2(z) = \frac{n^2 - 2n}{4},$$
and now Lemma 2.0.19 gives
$$g_2(E) = \frac{3n - 4}{4} - \frac{o_2(y)}{2}.$$
Also,
$$o_1(y) = 3n + 2 - 2(n-1) - \frac{n+2}{2} - \frac{n}{2} = 3.$$
Let y have shape (a, b, c), so $a \geq b \geq c$. By Corollary 2.0.17 (with $x_1 = x$, $x_2 = z$ and $x_3 = y$), we have $\gcd(a, b, c) = 1$. In particular, exactly one of a, b, c is even, and $c < \frac{n}{3}$. By Lemma 5.0.38 we have
$$o_2(y) = \frac{n-2}{2} + \gcd(a, b) + \gcd(a, c) + \gcd(b, c).$$

If c divides neither a nor b then
$$o_2(y) \leq \frac{n-2}{2} + \frac{n-c}{2} + 2\frac{c}{2} = n - 1 + \frac{c}{2} < \frac{7n-6}{6},$$
so
$$g_2(E) > \frac{n-3}{6} > 2.$$
Say c divides exactly one of a, b. Then c is odd, and one of $\gcd(a, c)$ and $\gcd(b, c)$ is at most $\frac{c}{3}$, so
$$o_2(y) \leq \frac{n-2}{2} + \frac{n-c}{2} + c + \frac{c}{3} = n - 1 + \frac{5c}{6} < \frac{23n - 18}{18}.$$
This gives
$$g_2(E) > \frac{2n-9}{18},$$
so $g_2(E) > 2$ when $n > 22$. If $n = 22$ then direct inspection shows that $\gcd(a, b) = \gcd(a, c) = 1$ and it follows easily that $g_2(E) > 2$.

If c divides both of a, b then $c = 1$ and
$$o_2(y) < \frac{n-2}{2} + \frac{n-2}{2} + 2 = n,$$
so
$$g_2(E) > \frac{n-4}{4} > 2.$$

We now assume that $f_1(u) \leq 1$ for all $u \in A$. We assume that $o_1(x) \geq o_1(y) \geq o_1(z)$. Since $g_1(E) = 0$, we have
$$o_1(x) + o_1(y) + o_1(z) = 3n + 2 - 2(n-1) = n + 4.$$
Therefore,
$$o_1(x) \geq \frac{n+4}{3}.$$
Let $d = |x|$. Assume $f_1(x) = 1$. Then $c_2(x) \geq 2$. Assume $d > 2$. Then by Lemma 5.0.40(7e), we have
$$\varepsilon_2(x) \geq \frac{n-3}{2n(n-1)}.$$
If $B = \emptyset$ then we have
$$g_2(E) \geq \frac{n-3}{8} > 2.$$
So, assume that $B \neq \emptyset$. Then n is even, and since $c_1(x) = 1$ we cannot have $d \in \{4, 8\}$. Say $d = 6$, so x has shape $1^1 2^a 3^b 6^c$. Since n is even we have $b > 0$. Now Lemma 5.0.40(7d), with $1 \leq a \leq \frac{n-4}{2}$ and $1 \leq b \leq \frac{n-3}{3}$, gives
$$\varepsilon_2(x) \geq \frac{4n-9}{3n(n-1)},$$
so
$$g_2(E) \geq \frac{5n-16}{24} > 2.$$
If $d \geq 10$ then Lemma 5.0.40(7f) gives
$$\varepsilon_2(x) \geq \frac{8n-40}{5n(n-1)},$$
so
$$g_2(E) \geq \frac{11n-80}{40} > 2.$$

Now assume that $d = 2$, so $\varepsilon_2(x) = 0$ and $o_1(x) = \frac{n+1}{2}$. This means that $o_1(y) + o_1(z) = \frac{n+7}{2}$, so
$$o_1(y) \geq \frac{n+7}{4}.$$
Moreover, n is odd, so $B = \emptyset$.

Let $f = |y|$. First assume that $f_1(y) = 1$. Then $c_2(y) + c_3(y) > 0$. Say $c_2(y) = 0$. If $f = 3$ then Lemma 5.0.40(7a) gives $\varepsilon_2(y) = \frac{2}{3n}$, so
$$g_2(E) \geq \frac{n-1}{6} > 2.$$
If $f \geq 6$ then Lemma 5.0.40(7e) gives
$$\varepsilon_2(y) \geq \frac{n-4}{2n(n-1)},$$
so
$$g_2(E) \geq \frac{n-4}{8} > 2.$$
Now say $c_2(y) > 0$. If $f > 2$ then the arguments which were used in the case $f_1(x) = 1$, $d > 2$, $B = \emptyset$ also apply here.

We now examine the case $f = 2$. In this case we have
- $o_1(z) = 3$,
- $i_2(s) = i_2(t) = n - 2$, and
- $i_2(x) = i_2(y) = \frac{1}{2}\left[\binom{n}{2} - \frac{n-1}{2}\right]$.

Direct calculation shows that
$$g_2(E) = \frac{3n-3}{4} - \frac{o_2(z)}{2}.$$
Let z have shape (a, b, c). We argue as we did in the case above where $f_1(x) = 2$ and $o_1(y) = 3$. We have
$$o_2(z) \leq \frac{n-1}{2} + \gcd(a,b) + \gcd(a,c) + \gcd(b,c).$$
By Corollary 2.0.17, we have $\gcd(a, b, c) = 1$, so $c < \frac{n}{3}$. If neither a nor b is a multiple of c then
$$o_2(z) \leq \frac{n-1}{2} + c + \frac{n-c}{2} \leq \frac{7n-3}{6}$$
and
$$g_2(E) \geq \frac{n-3}{6} > 2.$$
If exactly one of a, b is a multiple of c then we cannot have $a = b$. Therefore, if b does not divide a then $\gcd(a,b) \leq \frac{b}{2} \leq \frac{n-c}{4}$ while if $b | a$ then $b < a$, which forces $b \leq \frac{n-c}{3}$. Therefore,
$$o_2(z) \leq \frac{n-1}{2} + \frac{3c}{2} + \frac{n-c}{3} \leq \frac{22n-9}{18}$$
and
$$g_2(E) \geq \frac{5n-18}{36} > 2.$$
If c divides both a and b then $c = 1$ and
$$o_2(z) \leq \frac{n-1}{2} + 2 + \frac{n-c}{2} = n+1,$$

so
$$g_2(E) \geq \frac{n-5}{4} > 2.$$

We now assume that $f_1(y) = 0$. In this case we have $c_2(y) + c_3(y) \geq 3$. Let $f = |y|$. We first handle the case $f = 6$, so y has shape $2^a 3^b 6^c$. Since n is odd we have $b > 0$. If $a > 0$ then Lemma 5.0.40(8d) gives
$$\varepsilon_2(y) \geq \frac{n-3}{n(n-1)},$$
so
$$g_2(E) \geq \frac{n-3}{4} > 2.$$

If $a = 0$ then $b \geq 2$ and Lemma 5.0.40(8b) gives
$$\varepsilon_2(y) \geq \frac{5n-30}{6n(n-1)},$$
so
$$g_2(E) \geq \frac{5n-30}{24} > 2.$$

Since n is odd, we may now assume that if $c_2(y) > 0$ then $f \geq 10$. Say $c_2(y) \geq 2$. Then by Lemma 5.0.40(8f) we have
$$\varepsilon_2(y) \geq \left(\frac{1}{2} - \frac{1}{10}\right)\frac{3n-12}{n(n-1)} - \frac{n-3}{2n(n-1)} = \frac{7n-33}{10n(n-1)},$$
so
$$g_2(E) \geq \frac{7n-33}{40} > 2.$$

Say $c_2(y) = 1$, so $c_3(y) > 1$. We may assume $f \geq 12$. Now Lemma 5.0.40(8h) gives
$$\varepsilon_2(y) \geq \frac{17n-56}{30n(n-1)} + \frac{4n-20}{12n(n-1)} = \frac{27n-106}{30n(n-1)},$$
so
$$g_2(E) \geq \frac{27n-106}{120} > 2.$$

Say $c_2(y) = 0$, so $c_3(y) \geq 3$. Note that in this case we have
$$\sum_{2|m} c_m(y) \leq \frac{n-9}{4}.$$

If $f = 9$ then Lemma 5.0.40(8c) gives
$$\varepsilon_2(y) \geq \frac{2n-18}{n(n-1)},$$
so
$$g_2(E) \geq \frac{n-9}{2} > 2.$$

If $f \geq 12$ then Lemma 5.0.40(8e) gives
$$\varepsilon_2(y) \geq \left(\frac{1}{3} - \frac{1}{12}\right)\frac{2n-6}{n(n-1)} - \frac{n-9}{4n(n-1)} = \frac{n+3}{4n(n-1)},$$
so
$$g_2(E) \geq \frac{n+3}{16}.$$

This gives $g_2(E) > 2$ when $n > 29$. If $n \leq 29$, the conditions $|y| \geq 12$ and $o_1(y) \geq \frac{n+7}{4}$ force one of the conditions in the table below to hold.

$shape(y)$	$\varepsilon_2(y)$	Lower bound on $g_2(E)$
$3^7 4^1$	$17/300$	9
$3^7 4^2$	$17/203$	17
$3^8 5^1$	$12/203$	12

If $|y| = 3$ then $\varepsilon_2(y) = 0$ and $o_1(y) = \frac{n}{3}$. Also, $n \equiv 3 \bmod 6$,

$$o_1(z) = \frac{n+21}{6}$$

and

$$g_2(E) = \frac{n(n-1)}{4}\varepsilon_2(z)$$

Let $h = |z|$. First assume that $f_1(z) = 1$, and let s be the length of the shortest cycle in z other than the unique cycle of length one. Then $s \leq 5$. If $h = s$ then our assumptions on n and z force $s \geq 4$. In this case Lemma 5.0.40(7a) gives $\varepsilon_2(z) \geq \frac{1}{2n}$, so

$$g_2(E) \geq \frac{n-1}{8} > 2.$$

If $h \geq 2s$ then Lemma 5.0.40(7e) gives

$$\varepsilon_2(z) \geq \frac{n-1-s}{2n(n-1)} \geq \frac{n-6}{2n(n-1)},$$

so

$$g_2(E) \geq \frac{n-6}{8}.$$

This gives $g_2(E) > 2$ when $n > 21$. If $n = 21$ then $s \leq 3$ and it follows that

$$g_2(E) \geq \frac{n-4}{8} > 2.$$

Now assume that $f_1(z) = 0$. Let s be the length of the shortest cycle in z. Then $s \leq 5$. If $h = s$ then Proposition 2.0.16 forces $s = 5$. Now we have $\frac{n}{5} = \frac{n+21}{6}$, so $n = 105$. This is impossible by Corollary 2.0.18. If $h = 2s$ then Proposition 2.0.16 again gives $s = 5$, so z has shape $5^a 10^b$ with $a, b > 0$. Simultaneously solving the equations $5a + 10b = n$ and $a + b = \frac{n+21}{6}$ gives $a = \frac{2n+105}{15}$ and $b = \frac{n-105}{30}$. Since b is a positive integer, we have $n \geq 135$. Lemma 5.0.40(8b) gives

$$\varepsilon_2(z) = \frac{n^2 - 54n - 5355}{45n(n-1)},$$

so

$$g_2(E) \geq \frac{n^2 - 54n - 5355}{180} > 2.$$

Now assume that $h \geq 3s$. Note that since n is odd we have

$$\sum_{2|m} c_m(z) \leq \frac{n+15}{6}.$$

Now Lemma 5.0.40(8e) gives

$$\varepsilon_2(z) \geq \frac{2(s-1)(n-s)}{3sn(n-1)} - \frac{n+15}{6n(n-1)}.$$

If $s \geq 3$ then
$$\varepsilon_2(z) \geq \frac{5n - 69}{18n(n-1)},$$
so
$$g_2(E) \geq \frac{5n - 69}{72}.$$
It follows that $g_2(E) > 2$ whenever $n \geq 45$. Under our current assumptions, we need only check the cases where $n \in \{21, 27, 33, 39\}$. The conditions $c_1(z) = c_2(z) = 0$, $h \neq 3$ and $o_1(z) = \frac{n+21}{6}$ force $n > 21$, and either $c_3(z) \geq 2$ or z has shape $3^1 4^9$. In either case, we have $h \geq 12$ (as $h \neq 9$ by Proposition 2.0.16), and now Lemma 5.0.40(8k,8l) gives
$$\varepsilon_2(z) \geq \frac{5n - 36}{6n(n-1)}$$
and
$$g_2(E) \geq \frac{5n - 36}{24} > 2.$$
Now assume $s = 2$. If $h = 6$ then, since n is odd, we have $c_3(z) > 0$. Lemma 5.0.40(8d) gives
$$\varepsilon_2(z) \geq \frac{n - 3}{n(n-1)},$$
so
$$g_2(E) \geq \frac{n - 3}{4} > 2.$$
So, we may assume that $h \geq 10$. If $c_2(z) \geq 2$ then Lemma 5.0.40(8f) gives
$$\varepsilon_2(z) \geq \left(\frac{1}{2} - \frac{1}{10}\right)\frac{3n - 12}{n(n-1)} - \frac{n + 15}{6n(n-1)} = \frac{31n - 219}{30n(n-1)},$$
so
$$g_2(E) \geq \frac{31n - 219}{120} > 2.$$
Say $c_2(z) = 1$. Then $c_3(z) + c_4(z) + c_5(z) > 0$. If $c_3(z) > 0$ then Lemma 5.0.40(8h) (using $h \geq 12$) gives
$$\varepsilon_2(z) \geq \frac{5n - 35}{6n(n-1)}$$
and
$$g_2(E) \geq \frac{5n - 35}{24} > 2.$$
If $c_3(z) = 0$ and $c_4(z) > 0$ then Lemma 5.0.40(8i) (using $h \geq 20$) gives
$$\varepsilon_2(z) \geq \frac{13n - 108}{12n(n-1)}$$
and
$$g_2(E) \geq \frac{13n - 108}{48} > 2.$$
If $c_3(z) = c_4(z) = 0$ then $n \geq 87$. Lemma 5.0.40(8j) gives
$$\varepsilon_2(z) \geq \frac{13n - 117}{30n(n-1)}$$
and
$$g_2(E) \geq \frac{13n - 117}{120} > 2.$$

Now assume that $f_1(x) = 0$. In this case we have $c_2(x) \geq 4$. We show first that if $|x| > 6$ then
$$\sum_{2|m} c_m(x) \leq \frac{n-5}{2}.$$
Indeed, if $c_m(x) > 0$ for some $m \geq 7$ then the claim holds with no restrictions on $|x|$. The same is true if $c_5(x) > 0$. If $c_m(x) = 0$ for all $m \notin \{2,3,4,6\}$ then the condition $|x| > 6$ forces $c_4(x) > 0$ and $c_3(x) + c_6(x) > 0$ and the claim holds.

If $|x| > 2$ then $o_1(x) < \frac{n}{2}$, and our assumption that $o_1(x) \geq o_1(y) \geq o_1(z)$ guarantees that neither of y, z is a fixed-point-free involution. It then follows from Proposition 2.0.16 and Lemma 5.0.40(8a,9) that
$$\varepsilon_2(y) + \varepsilon_2(z) \geq -\frac{1}{4(n-1)}.$$

If $|x| \geq 10$ then Lemma 5.0.40(8f) gives
$$\varepsilon_2(x) \geq \left(\frac{1}{2} - \frac{1}{10}\right)\frac{3n-12}{n(n-1)} - \frac{n-5}{2n(n-1)} = \frac{7n-23}{10n(n-1)}.$$
Therefore,
$$g_2(E) \geq \frac{9n-46}{80}.$$
This gives $g_2(E) > 2$ when $n > 22$. If $n \in \{21, 22\}$ then by Lemma 5.0.40(8a,9) we have
$$\varepsilon_2(y) + \varepsilon_2(z) \geq -\frac{1}{22(n-1)}$$
and
$$g_2(E) \geq \frac{72n-253}{440} > 2.$$

If $|x| = 8$ then $B = \emptyset$ by Proposition 2.0.16, and $c_8(x) > 0$. Lemma 5.0.40(8f) gives
$$\varepsilon_2(x) \geq \left(\frac{1}{2} - \frac{1}{8}\right)\frac{3n-12}{n(n-1)} - \frac{n-6}{2n(n-1)} = \frac{5n-12}{8n(n-1)},$$
so
$$g_2(E) \geq \frac{5n-12}{32} > 2.$$

Say $|x| = 6$, so x has shape $2^a 3^b 6^c$ with $4 \leq a \leq \frac{n-3}{2}$. Lemma 5.0.40(8d) gives
$$\varepsilon_2(x) = \frac{2a(n-1) - 4a^2}{3n(n-1)} + \frac{2b(n-1) - b(6b-4)}{4n(n-1)} - \frac{1}{6(n-1)}.$$
Note that the second summand on the right side of the equality just stated is always nonnegative. If we fix n and b then $\varepsilon_2(x)$ is a quadratic function of a with negative leading coefficient. Comparing $\varepsilon_2(x)$ when $(a,b) = (4,0)$ and $(\frac{n-6}{2}, 0)$ shows that if $4 \leq a \leq \frac{n-6}{2}$ then
$$\varepsilon_2(x) \geq \frac{3n-20}{2n(n-1)}.$$
In this case, we have
$$g_2(E) \geq \frac{5n-40}{16} > 2.$$
If it is not the case that $4 \leq a \leq \frac{n-6}{2}$ then x has shape $2^{\frac{n-3}{2}} 3^1$ and
$$\varepsilon_2(x) = \frac{n-3}{n(n-1)}.$$

In this case, n is odd so $B = \emptyset$. Therefore
$$g_2(E) \geq \frac{n-3}{4} > 2.$$
If $|x| = 4$ then $B = \emptyset$ by Proposition 2.0.16. Also, x has shape $2^a 4^b$, with $4 \leq a \leq \frac{n-4}{2}$. By Lemma 5.0.40(8b) we have
$$\varepsilon_2(x) = \frac{(2a-1)(n-2a) - 4a}{4n(n-1)}.$$
If we fix n then $\varepsilon_2(x)$ is a quadratic function of a with negative leading coefficient. Comparing $\varepsilon_2(x)$ when $a = 4$ and $a = \frac{n-4}{2}$ gives
$$\varepsilon_2(x) \geq \frac{n-6}{2n(n-1)}.$$
Therefore
$$g_2(E) \geq \frac{n-6}{8}.$$
This gives $g_2(E) > 2$ when $n > 22$. If $n = 22$ then one of y, z does not have shape 11^2 by Proposition 2.0.16, and it follows from Lemma 5.0.40(9) that $\varepsilon_2(y) + \varepsilon_2(z) > 0$ and $g_2(E) > 2$.

Now assume that $|x| = 2$. Then n is even and $\varepsilon_2(x) = -\frac{1}{2(n-1)}$, so
$$(5.11) \qquad g_2(E) = \frac{n(n-1)}{4}(\varepsilon_2(y) + \varepsilon_2(z)) - \frac{n}{8}.$$
By Proposition 2.0.16 and Lemma 5.0.40(9), both $\varepsilon_2(y)$ and $\varepsilon_2(z)$ are nonnegative. Also,
$$o_1(y) + o_1(z) = n + 4 - \frac{n}{2} = \frac{n+8}{2},$$
so we have
$$o_1(y) \geq \frac{n+8}{4}.$$
Say $f_1(y) = 1$. Then $c_2(y) + c_3(y) \geq 3$. Since n is even, we see that y has some odd cycle other than its fixed point. Thus either $|y| = 3$ or $|y| \geq 6$. If $c_2(y) \geq 2$ then Lemma 5.0.40(7f) gives
$$\varepsilon_2(y) \geq \frac{4n-20}{3n(n-1)},$$
so
$$g_2(E) \geq \frac{5n-40}{24} > 2.$$
If $c_2(y) = 1$ then $c_3(y) \geq 1$. Again we have $|y| \geq 6$, and if $|y| > 6$ then Lemma 5.0.40(7g) gives
$$\varepsilon_2(y) \geq \frac{2n-9}{n(n-1)} - \frac{n-6}{2n(n-1)} = \frac{3n-12}{2n(n-1)}.$$
Therefore,
$$g_2(E) \geq \frac{n-6}{4} > 2.$$
If $|y| = 6$ then, since $c_2(y) = 1$, Lemma 5.0.40(7d) gives
$$\varepsilon_2(y) \geq \frac{4n-12}{3n(n-1)},$$

so
$$g_2(E) \geq \frac{5n-24}{24} > 2.$$
If $c_2(y) = 0$ then $c_3(y) \geq 3$. If $|y| \geq 12$ then Lemma 5.0.40(7j) gives
$$\varepsilon_2(y) \geq \frac{3n-21}{2n(n-1)},$$
so
$$g_2(E) \geq \frac{2n-21}{8} > 2,$$
If $|y| = 9$ then y has shape $1^1 3^a 9^b$, with $3 \leq a \leq \frac{n-10}{3}$. Lemma 5.0.40(7c) gives
$$\varepsilon_2(y) = \frac{8(n-1) + 6a(n-3a-2)}{9n(n-1)}.$$
If we fix n, $\varepsilon_2(y)$ is quadratic in a with negative leading coefficient and is therefore minimized at $a = 3$ or at $a = \frac{n-10}{3}$. Comparing $\varepsilon_2(y)$ at these two values of a gives
$$\varepsilon_2(y) \geq \frac{8n-56}{3n(n-1)},$$
so
$$g_2(E) \geq \frac{13n-112}{24} > 2.$$
If $|y| = 6$ then y has shape $1^1 3^a 6^b$ with $3 \leq a \leq \frac{n-7}{3}$. Lemma 5.0.40(7b) gives
$$\varepsilon_2(y) = \frac{3a+4}{6n} - \frac{3a^2}{2n(n-1)}.$$
Arguing as we did in the case $|y| = 9$, we see that $\varepsilon_2(y)$ is minimized when $a = \frac{n-7}{3}$, giving
$$\varepsilon_2(y) \geq \frac{5n-23}{3n(n-1)}.$$
Therefore,
$$g_2(E) \geq \frac{7n-46}{24} > 2.$$
If $|y| = 3$ then Lemma 5.0.40(7a) gives
$$\varepsilon_2(y) = \frac{2}{3n},$$
so
$$g_2(E) = \frac{n-4}{24} + \frac{n(n-1)}{4} \varepsilon_2(z).$$
This gives $g_2(E) > 2$ when $n > 52$. Also, the Riemann-Hurwitz formula gives
$$o_1(z) = \frac{n+20}{6}.$$
If follows that if $n \leq 52$ then
$$\sum_{j=1}^{4} c_j(z) \geq 2.$$
If $f_1(z) = 1$ then Lemma 5.0.40(7a,7e) and the fact that n is even give
$$\varepsilon_2(z) \geq \frac{n-4}{2n(n-1)}$$

5. ACTIONS ON 2-SETS

(this bound arising in (7e) with $s = 3$ and $|z| = 6$), so

$$g_2(E) \geq \frac{n-4}{6} > 2.$$

If $f_1(z) = 0$ then, since $\sum_{2|m} c_m(z) \leq \frac{n+8}{6}$, Lemma 5.0.40(8f,8h,8i,8k,8l,8n) gives

$$\varepsilon_2(z) \geq \frac{17n - 65}{30n(n-1)},$$

this bound arising from taking $d = 6$ in (8h). It follows that

$$g_2(E) \geq \frac{22n - 85}{120} > 2.$$

We may now assume that $f_1(y) = 0$. By Proposition 2.0.16, we have $|y| \neq 4$. It then follows from our assumption $o_1(y) \geq \frac{n+8}{4}$ that $c_2(y) + c_3(y) \geq 5$.

We begin with the case $|y| = 6$. If $c_2(y) = 0$ then y has shape $3^a 6^b$ with $5 \leq a \leq \frac{n-6}{3}$ and Lemma 5.0.40(8b) gives

$$\varepsilon_2(y) = \frac{(3a-1)(n-3a)}{6n(n-1)}.$$

For fixed n, $\varepsilon_2(y)$ is minimized when $a = 5$ or when $a = \frac{n-6}{3}$. Comparing $\varepsilon_2(y)$ for these two values of a gives

$$\varepsilon_2(y) \geq \frac{n-7}{n(n-1)},$$

so

$$g_2(E) \geq \frac{n-14}{8}.$$

In addition, we have

$$\frac{n-3}{3} \geq o_1(y) \geq \frac{n+8}{4},$$

and it follows that $n \geq 36$, so $g_2(E) > 2$. Now assume that y has shape $2^a 3^b 6^c$ with $a > 0$. By Proposition 2.0.16 we have $b > 0$, and since n is even we have $b \geq 2$. By Lemma 5.0.40(8d) we have

$$\varepsilon_2(y) = \frac{2a}{3n} - \frac{4a^2}{3n(n-1)} + \frac{b}{2n} - \frac{b(6b-4)}{4n(n-1)} - \frac{1}{6(n-1)}.$$

We have $1 \leq a \leq \frac{n-3}{2}$ and $2 \leq b \leq \frac{n-2}{3}$. It follows that

$$\varepsilon_2(y) \geq \frac{2n-6}{3n(n-1)} + \frac{n-2}{2n(n-1)} - \frac{1}{6(n-1)} = \frac{n-3}{n(n-1)}.$$

Equation (5.11) gives

$$g_2(E) \geq \frac{n-6}{8},$$

So $g_2(E) > 2$ when $n > 22$. If $n = 22$ then $b \leq \frac{n-4}{3}$ and it follows that

$$\varepsilon_2(y) \geq \frac{4n-16}{3n(n-1)}$$

and

$$g_2(E) \geq \frac{5n-32}{24} > 2.$$

We now assume that $|y| \neq 6$. Say $c_2(y) > 0$ and $c_3(y) > 0$. Then $|y| \geq 12$. If $c_2(y) \geq 2$ then Lemma 5.0.40(8g) gives

$$\varepsilon_2(y) \geq \frac{4n-15}{3n(n-1)} + \frac{n-5}{3n(n-1)} - \frac{n-6}{2n(n-1)} = \frac{7n-22}{6n(n-1)},$$

so

$$g_2(E) \geq \frac{2n-11}{12} > 2.$$

Say $c_2(y) = 1$. Since $n > 20$ is even and $o_1(y) \geq \frac{n+8}{4}$, either y has shape $2^1 3^6 4^1$ or $n \geq 26$. In the first case, we have $\varepsilon_2(y) = \frac{2}{23}$ and $g_2(E) \geq 9$. So, assume $n \geq 26$. If $|y| = 12$ then, in the language of Lemma 5.0.40(8h) we have $J_3 = \{2\}$ and

$$\varepsilon_2(y) \geq \frac{3n-13}{3n(n-1)}.$$

This gives

$$g_2(E) \geq \frac{3n-26}{24} > 2.$$

If $|y| \geq 18$ then, since $c_3(y) \geq 4$ we have

$$\sum_{m \in J_3} c_m(y) \leq 1 + \frac{n-14}{10} = \frac{n-4}{10}$$

and

$$\varepsilon_2(y) \geq \frac{91n-314}{90n(n-1)}.$$

This gives

$$g_2(E) \geq \frac{46n-314}{360} > 2.$$

Say $c_3(y) = 0$. By Lemma 5.0.42, y has at least two cycles of odd length, and we cannot have $|y| \in \{8, 12, 16\}$. If $|y| \geq 18$ then Lemma 5.0.40(8f), we have

$$\varepsilon_2(y) \geq \frac{12n-48}{9n(n-1)} - \frac{n-10}{2n(n-1)} = \frac{5n-2}{6n(n-1)}$$

and

$$g_2(E) \geq \frac{n-1}{12}.$$

This gives $g_2(E) > 2$ when $n > 24$. Moreover, we cannot have $n = 22$, and if $n = 24$ then y has shape $2^5 4^1 5^2$ or $2^6 5^1 7^1$. In either case, direct calculation gives $g_2(E) > 2$. If $|y| = 14$ then Lemma 5.0.40(8f) gives

$$\varepsilon_2(y) \geq \frac{11n+26}{14n(n-1)}$$

and

$$g_2(E) \geq \frac{2n+13}{28} > 2.$$

If $|y| = 10$ then Lemma 5.0.40(8f) gives

$$\varepsilon_2(y) \geq \frac{7n+2}{10n(n-1)}$$

and

$$g_2(E) \geq \frac{n+1}{20}.$$

This gives $g_2(E) > 2$ when $n \geq 40$. If $n \leq 38$, the condition $o_1(y) \geq \frac{n+8}{4}$ forces the shape of y to be one of $2^9 5^2 10^1$, $2^8 5^2 10^1$ or $2^a 5^{(n-2a)/5}$ with $5 \leq a \leq \frac{n-10}{2}$.

In the first two cases, direct calculation gives $g_2(E) > 2$. In the last case, direct calculation shows that
$$\varepsilon_2(y) = \frac{a(n-2a-1)}{n(n-1)} \geq \frac{9n-90}{2n(n-1)},$$
so
$$g_2(E) \geq \frac{4n-45}{4} > 2.$$

Finally, say $c_2(y) = 0$. Assume first that also $|y| > 3$. Then, since $o_1(y) \geq \frac{n+8}{4}$ and n is even, either y has shape $3^8 4^1$ or $n > 30$. In the first case, direct calculation gives $g_2(E) > 2$. In the second case, we have $|y| \geq 9$ and Lemma 5.0.40(8k) gives
$$\varepsilon_2(y) \geq \frac{10n-60}{9n(n-1)} - \frac{n-15}{4n(n-1)} = \frac{31n-105}{36n(n-1)},$$
so
$$g_2(E) \geq \frac{13n-105}{144} > 2.$$

We are left with the case $|y| = 3$. In this case, we have $\varepsilon_2(y) = 0$ and
$$g_2(E) = \frac{n(n-1)}{4}\varepsilon_2(z) - \frac{n}{8}.$$
Also,
$$o_1(z) = \frac{n+24}{6}.$$

Say $f_1(z) = 1$. Let s be the length of the shortest nontrivial cycle in z. Then $s \leq 5$, and if $s > 2$ we get a lower bound on n, as described in the following table.

s	3	4	5
$n \geq$	24	42	96

If $|z| = 5$ then $n = 96$ and direct calculation gives $g_2(E) = 7$. If z has shape $1^1 5^a 10^b$, then from the equations $1 + a + b = \frac{n+24}{6}$ and $1 + 5a + 10b = n$ we get $a = \frac{2n+93}{15}$. Lemma 5.0.40(7b) gives
$$\varepsilon_2(z) = \frac{2n^2 - 27n - 9000}{90n(n-1)},$$
so
$$g_2(E) = \frac{2n^2 - 27n - 9000}{360} > 2.$$

If $|z| \notin \{5, 10\}$ and $s > 2$ then, using Lemma 5.0.40(7e) we see that one of the cases listed in the following table occurs. Note that if $s = 4$ then $|z| \geq 20$ and if $s = 3$ then $|z| \geq 12$.

s	lower bound on $\varepsilon_2(z)$	lower bound on $g_2(E)$
5	$\frac{2n-12}{3n(n-1)}$	$\frac{n-24}{24}$
4	$\frac{4n-20}{5n(n-1)}$	$\frac{3n-40}{40}$
3	$\frac{3n-12}{4n(n-1)}$	$\frac{n-12}{16}$

Given our lower bounds on n determined above, it follows that $g_2(E) > 2$ unless $s = 3$ and $n \leq 42$ (as $6|n$). Note that if $s = 3$ we cannot have $|z| \in \{6, 9\}$ (again because $6|n$). Thus if $c_3(z) \geq 2$ then Lemma 5.0.40(7j) gives
$$\varepsilon_2(z) \geq \frac{3n-21}{2n(n-1)}$$

and
$$g_2(E) \geq \frac{2n - 21}{8} > 2.$$

If $c_3(z) = 1$ then, under the current conditions, the shape of z is one of $1^1 3^1 4^8$, $1^1 3^1 4^8 6^1$ or $1^1 3^1 4^7 5^2$. In each case, direct calculation gives $g_2(E) > 2$.

If $s = 2$, let t be the length of the second shortest nontrivial cycle of z. Then one of the cases in the following table occurs. Note that if $t = 2$ then $|z| \geq 6$ and if $t = 4$ then $|z| \geq 20$.

t	lower bound on $\varepsilon_2(z)$	lower bound on $g_2(E)$	lemma used
5	$\frac{n-3}{n(n-1)}$	$\frac{n-6}{8}$	5.0.40(7i)
4	$\frac{17n-99}{10n(n-1)}$	$\frac{12n-99}{40}$	5.0.40(7h)
3	$\frac{n-3}{n(n-1)}$	$\frac{n-6}{8}$	5.0.40(7g)
2	$\frac{4n-20}{3n(n-1)}$	$\frac{5n-40}{24}$	5.0.40(7f)

Since $n \geq 24$, we have $g_2(E) > 2$ in each case listed in the table.

Now assume $f_1(z) = 0$. By Proposition 2.0.16 we have $|z| \notin \{2, 3, 4, 8, 9\}$. If $|z| = 7$ then $\frac{n+24}{6} = \frac{n}{7}$, which is impossible. If $|z| = 5$ then we have $\frac{n+24}{6} = \frac{n}{5}$, so $n = 120$. Also, $\varepsilon_2(z) = 0$ and it follows that $g_2(E) = -15$, which is also impossible. Say $|z| = 6$, so z has shape $2^a 3^b 6^c$. By Proposition 2.0.16 we have $a > 0$ and $b > 0$. Now the equations $a + b + c = \frac{n+24}{6}$ and $2a + 3b + 6c = n$ give $4a + 3b = 24$, and it follows that $(a, b) = (3, 4)$. Now Lemma 5.0.40(8d) gives
$$\varepsilon_2(z) = \frac{23n - 216}{6n(n-1)}$$
and
$$g_2(E) = \frac{5n - 54}{6} > 2.$$

We now assume that $|z| \geq 10$. Let (s, t) be the lengths of the two shortest cycles in z. Then one of the cases listed in the table below occurs. Note that z has at least two cycles of odd length, so in all cases we have
$$\sum_{2 \mid m} c_m(z) \leq \frac{n+12}{6}.$$

(s,t)	lower bound on n	lower bound on $\varepsilon_2(z)$	lower bound on $g_2(E)$	lemma used	comments
$(2,2)$	24	$\frac{31n-204}{30n(n-1)}$	$\frac{16n-204}{120}$	5.0.40(8f, 8g)	
$(2,3)$	24	$\frac{27n-106}{30n(n-1)}$	$\frac{6n-53}{60}$	5.0.40(8h)	$\sum_{m \in J_3} c_m(x) \leq \frac{n+2}{10}$
$(2,4)$	48	$\frac{13n-102}{12n(n-1)}$	$\frac{7n-102}{48}$	5.0.40(8i)	$|z| \geq 20$
$(2,5)$	102	$\frac{3n-12}{5n(n-1)}$	$\frac{n-24}{40}$	5.0.40(8j)	$|z| = 10 \Rightarrow \sum_{m \in J_5} c_m(x) = 1$
$(3,3)$	30	$\frac{13n-114}{12n(n-1)}$	$\frac{7n-114}{48}$	5.0.40(8k)	
$(3,4)$	48	$\frac{7n-47}{6n(n-1)}$	$\frac{4n-47}{24}$	5.0.40(8l)	$|z| \geq 36$
$(3,5)$	108	$\frac{27n-188}{30n(n-1)}$	$\frac{3n-47}{30}$	5.0.40(8m)	
$(4,4)$	54	$\frac{37n-396}{30n(n-1)}$	$\frac{11n-198}{60}$	5.0.40(8n)	$|z| \geq 20$
$(4,5)$	114	$\frac{31n-222}{30n(n-1)}$	$\frac{8n-111}{60}$	5.0.40(8o)	
$(5,5)$	126	$\frac{11n+15}{15n(n-1)}$	$\frac{7n+330}{120}$	5.0.40(8p)	$c_5(z) \geq 24$

It follows that $g_2(E) > 2$ whenever (s,t,n) is not one of $(2,2,24)$, $(2,3,24)$, $(2,5,102)$ or $(3,3,30)$. Say $(s,t,n) = (2,2,24)$. If $c_3(z) > 0$ then Lemma 5.0.40(8g) gives

$$\varepsilon_2(z) \geq \frac{9n-52}{6n(n-1)},$$

so

$$g_2(E) \geq \frac{3n-26}{12} > 2.$$

If $c_3(z) = 0$ then z has shape $2^6 5^1 7^1$ or $2^5 4^1 5^2$. If $(s,t,n) = (2,3,24)$ then z has shape $2^1 3^6 4^1$. If $(s,t,n) = (2,5,102)$ then z has shape $2^1 5^{20}$. If $(s,t,n) = (3,3,30)$ then z has shape $3^7 4^1 5^1$ or $3^6 4^3$. In all of the cases just listed, direct calculation gives $g_2(E) > 2$. □

LEMMA 5.0.50. *Theorem 4.0.30 is true if we assume that $|A| = 3$ and $|T| \geq 3$.*

PROOF. If $g_1(E) \geq 1$ then Equation (5.1) of Lemma 5.0.43 and Lemma 5.0.40(1-6,8a,9) give

$$\begin{aligned} g_2(E) - g_1(E) &\geq \frac{(r-3)(n-3)}{4} - \frac{3n}{8} \\ &\geq \frac{3n-18}{8}. \end{aligned}$$

If follows that $3g_2(E) > r$ and $g_2(E) - g_1(E) > 2$.

If $g_1(E) = 0$ then Lemmas 5.0.43, 5.0.40(1-6,8a,9) and 5.0.42 give

$$g_2(E) \geq \frac{(2r-6)(n-3) - 5n + 12}{8}.$$

It follows that if $r \geq 7$ then $3g_2(E) > r$ (so $g_2(E) > 2$). If $r = 6$ then $g_2(E) > 2$ (so $3g_2(E) > r$) whenever $n > 22$.

Assume that $r = 6$ and $n \leq 22$. Then $E = T \cup A$. Assume $A = \{x,y,z\}$. We have

$$g_2(E) = \frac{n-3}{4} + \frac{n(n-1)}{4}(\varepsilon_2(x) + \varepsilon_2(y) + \varepsilon_2(z)).$$

If $B = \emptyset$ then by Lemma 5.0.40(9) we have $g_2(E) > 2$.

By Lemma 5.0.42, we may now assume that $B = \{x\}$, so $n = 22$. Say $|x| = 22$. Then Lemma 5.0.40(8a,9) gives

$$g_2(E) \geq \frac{21n - 66}{88} > 2.$$

We are left with the case where $|x| = 2$. In this case we have

$$g_2(E) = \frac{n-6}{8} + \frac{n(n-1)}{4}(\varepsilon_2(y) + \varepsilon_2(z)).$$

Since $n = 22$, by Proposition 2.0.16 and Lemma 5.0.40(9), we have $\varepsilon_2(y) + \varepsilon_2(z) > 0$ and $g_2(E) > 2$. □

LEMMA 5.0.51. *Theorem 4.0.30 is true if we assume that $|A| \leq 2$.*

PROOF. If $g_1(E) > 0$ then by Equation (5.1) of Lemma 5.0.43 and Lemma 5.0.40(1,2,3,8a,9) we have
$$g_2(E) - g_1(E) \geq \frac{(r-2)(n-3)}{4} - \frac{n}{4}$$
$$\geq \frac{2n-9}{4}.$$
It follows that $3g_2(E) > r$ and $g_2(E) > 2$.

If $g_1(E) = 0$ then Lemmas 5.0.43, 5.0.40(1,2,3) and 5.0.42 give
$$g_2(E) \geq \frac{(2r-4)(n-3) - 5n + 12}{8}.$$
If follows that $3g_2(E) > r$ (so $g_2(E) > 2$) whenever $r \geq 6$. Say $r = 5$. If $E \neq T \cup A$ then Lemmas 5.0.43, 5.0.40(1-3) and 5.0.42 give
$$g_2(E) \geq -\frac{n-3}{2} + \frac{n-5}{2} + \frac{n-3}{2} - \frac{n}{8} = \frac{3n-20}{8} > 2.$$
Similarly, if B does not contain an involution then
$$g_2(E) \geq \frac{3n-12}{16}.$$
and $g_2(E) > 2$. If $E = T \cup A$ and B contains an involution then $\langle E \rangle$ is generated by three transpositions and a fixed-point-free involution. However, this contradicts our assumptions that $\langle E \rangle$ is transitive and $n > 20$. □

CHAPTER 6

Actions on 3-sets - the proof of Theorem 4.0.31

The proof of Theorem 4.0.31 is similar in spirit to that of Theorem 4.0.30. We begin with a lemma giving lower bounds of $\delta_{2,k}(x)$ for all k. These bounds will be of use in other parts of the paper. We then derive a much simpler analogue of Lemma 5.0.40, which suffices to prove Theorem 4.0.31 when condition (1) of Theorem 4.0.30 holds. Then explicit calculations will show that Theorem 4.0.31 is true when condition (2) of Theorem 4.0.30 holds.

LEMMA 6.0.52. *Let $x \in S_n$ and let $k \in \{2, \ldots, n-2\}$. Then the triple (x, n, k) satisfies one of the conditions*

(A) $\delta_{2,k}(x) \geq 0$, *or*
(B) $f_2(x) = 0$, *or*
(C) $k \in \{3, n-3\}$ *and x has shape $1^1 2^1 3^{(n-3)/3}$, or*
(D) $k = 4$ *and x has shape $1^2 3^2$.*

In any case,
$$\delta_{2,k}(x) \geq -\frac{2}{(n-1)(n-2)}.$$

PROOF. Say x has shape $(\lambda_1, \ldots, \lambda_r)$. If $\lambda_1 = 1$ then $\delta_{2,k}(x) = 0$ for all k, so we assume that $\lambda_1 \geq 2$. Write
$$x = \sigma_1 \ldots \sigma_r,$$
where the σ_i are disjoint cycles with $|supp(\sigma_i)| = \lambda_i$. If $\lambda_1 > 5$ then set
$$m = \max\{i \in [r] : \lambda_i > 5\}.$$
For each $i \in [m]$, choose some permutation τ_i of $supp(\sigma_i)$ such that each cycle in τ_i has length 3, 4 or 5. Set
$$y = \prod_{i=1}^{m} \tau_i \prod_{i=m+1}^{r} \sigma_i.$$
For any y obtained this way we have $f_2(y) = f_2(x)$ and $f_k(y) \geq f_k(x)$ for all k. This gives
$$\delta_{2,k}(y) \geq \delta_{2,k}(x)$$
for $2 \leq k \leq n-2$. Now if (y, n, k) satisfies one of conditions (A),(B) then (x, n, k) satisfies the same condition. If $(y, n, 3)$ satisfies condition (C) for every y which can be obtained from x in this manner then x has shape $1^1 2^1 3^c 6^f$ with $f > 0$. In this case we have
$$\frac{f_3(x)}{\binom{n}{3}} = \frac{2(n-6f)}{n(n-1)(n-2)} < \frac{2}{n(n-1)} = \frac{f_2(x)}{\binom{n}{2}},$$
so $(x, n, 3)$ satisfies condition (A). If $(y, 8, 4)$ satisfies condition (D) for every y which can be obtained from x in this manner then the shape of x is $1^2 6^1$. In this case

$f_4(x) = 0$ so $(x, 8, 4)$ satisfies condition (A). It now follows that we may assume that

- $2 \leq \lambda_1 \leq 5$.

We continue under this assumption.

We will prove that (x, n, k) satisfies one of (A)-(D) by induction on n. For $n \leq 5$ there is nothing to prove, so we assume that $n \geq 6$. Since $\lambda_1 < n$ we have $r > 1$. Let x^- be any element of $S_{n-\lambda_1}$ which has shape $(\lambda_2, \ldots, \lambda_r)$. If x fixes $S \subset [n]$ then
$$S \cap supp(\sigma_1) \in \{\emptyset, supp(\sigma_1)\}.$$
It follows that for any k we have
$$f_k(x) = f_k(x^-) + f_{k-\lambda_1}(x^-),$$
if we make the natural definition that for $g \in S_l$ we have $f_k(g) = 0$ when $k < 0$ or $k > l$.

The key fact in our inductive proof is the following one

(∗) Say $2 \leq k - \lambda_1 \leq k \leq (n - \lambda_1) - 2$. If $(x^-, n - \lambda_1, k)$ and $(x^-, n - \lambda_1, k - \lambda_1)$ both satisfy condition (A) then (x, n, k) also satisfies (A).

To prove (∗), we note that if the given assumptions are satisfied then

$$\frac{f_k(x)}{\binom{n}{k}} = \frac{\binom{n-\lambda_1}{k}}{\binom{n}{k}} \frac{f_k(x^-)}{\binom{n-\lambda_1}{k}} + \frac{\binom{n-\lambda_1}{k-\lambda_1}}{\binom{n}{k}} \frac{f_{k-\lambda_1}(x^-)}{\binom{n-\lambda_1}{k-\lambda_1}}$$

$$\leq \frac{f_2(x^-)}{\binom{n-\lambda_1}{2}} \left[\frac{\binom{n-\lambda_1}{k} + \binom{n-\lambda_1}{k-\lambda_1}}{\binom{n}{k}}\right]$$

$$\leq \frac{\binom{n}{2}}{\binom{n-\lambda_1}{2}} \frac{f_2(x)}{\binom{n}{2}} \left[\frac{\binom{n-\lambda_1}{k} + \binom{n-\lambda_1}{k-\lambda_1}}{\binom{n}{k}}\right].$$

Since $(x, n, 2)$ and $(x, n, n-2)$ certainly satisfy (A), it suffices to show that

(6.1) $$\frac{\binom{n-\lambda_1}{k} + \binom{n-\lambda_1}{k-\lambda_1}}{\binom{n}{k}} \leq \frac{(n-\lambda_1)(n-\lambda_1-1)}{n(n-1)}$$

for $3 \leq k \leq n - 3$. Straightforward computation shows that inequality (6.1) is equivalent to

(6.2) $$\prod_{i=0}^{\lambda_1-1}(n-k-i) + \prod_{i=0}^{\lambda_1-1}(k-i) \leq \prod_{j=2}^{\lambda_1+1}(n-j).$$

So, for nonnegative integers n, m, k with $k, m \leq n$, define
$$\phi(n, m, k) := \prod_{i=0}^{m-1}(n-k-i) + \prod_{i=0}^{m-1}(k-i).$$

Now if $k < n$ then

(6.3) $$\phi(n, m, k) - \phi(n, m, k+1) = m \left[\prod_{i=1}^{m-1}(n-k-i) - \prod_{i=0}^{m-2}(k-i)\right].$$

If we fix n and m then the term on the right side of equation (6.3) is a nonincreasing function of k. It follows that if n and λ_1 are fixed and $3 \leq k \leq n-3$ then $\phi(n, \lambda_1, k)$ is maximized at $k = 3$ (and at $k = n - 3$).

If $k = 3$ and $\lambda_1 \geq 4$ then
$$\prod_{i=0}^{\lambda_1-1}(3-i) = 0,$$
so
$$\begin{aligned}
\phi(n, \lambda_1, 3) &= \prod_{i=0}^{\lambda_1-1}(n-3-i) \\
&= \prod_{i=3}^{\lambda_1+2}(n-i) \\
&\leq \prod_{j=2}^{\lambda_1+1}(n-j)
\end{aligned}$$
and inequality (6.2) holds. If $k = \lambda_1 = 3$ then inequality (6.2) is equivalent to the inequality
$$(6.4) \qquad (n-2)(n-5) \geq 0,$$
while if $k = 3$ and $\lambda_1 = 2$ then (6.2) is equivalent to
$$(6.5) \qquad n \geq 6.$$
By assumption we have $n \geq 6$, so both inequalities (6.4) and (6.5) hold. This completes the proof that fact (∗) holds.

Certainly we cannot always apply fact (∗), since the assumptions given therein may not hold. We must examine the cases in which the assumptions do not hold, namely those cases where

- k is *too big*, that is, $k > n - \lambda_1 - 2$, or *too small*, that is, $k - \lambda_1 < 2$, or both too big and too small, or
- k is neither too big nor too small, but at least one of $(x^-, n - \lambda_1, k)$ or $(x^-, n - \lambda_1, k - \lambda_1)$ satisfies one of (B),(C) or (D).

Since there is nothing to prove if $k \in \{2, n-2\}$ we assume from now on that $3 \leq k \leq n-3$. The case $\lambda_1 = 2$ is slightly different then the case $3 \leq \lambda_1 \leq 5$, so we assume temporarily that $\lambda_1 = 2$. If k is too big then $k = n - 3$ while if k is too small then $k = 3$. Since $f_k(x) = f_{n-k}(x)$ for all k, we can prove the theorem in the case where k is too big or too small by showing that $(x, n, 3)$ satisfies condition (A). Now x has shape $1^a 2^b$ with $b = \frac{n-a}{2} > 0$. We have
$$\begin{aligned}
\frac{f_2(x)}{\binom{n}{2}} &= \frac{\binom{a}{2} + b}{\binom{n}{2}} \\
&= \frac{a(a-1) + n - a}{n(n-1)},
\end{aligned}$$
while
$$\begin{aligned}
\frac{f_3(x)}{\binom{n}{3}} &= \frac{\binom{a}{3} + ab}{\binom{n}{3}} \\
&= \frac{a(a-1)(a-2) + 3a(n-a)}{n(n-1)(n-2)}.
\end{aligned}$$

It follows that $\delta_{2,3}(x) \geq 0$ if and only if

(6.6) $$n + a^2 - 4a - 2 \geq 0.$$

Inequality (6.6) holds whenever $n \geq 6$, so if $k \in \{3, n-3\}$ then (x, n, k) satisfies condition (A) as claimed. We may now assume that k is neither too big nor too small. By observation, neither $(x^-, n-2, k)$ nor $(x^-, n-2, k-2)$ satisfies any of the conditions (B),(C),(D). By inductive hypothesis and fact $(*)$, (x, n, k) satisfies condition (A).

Now assume $3 \leq \lambda_1 \leq 5$. Say k is both too big and too small, so

- $n - (\lambda_1 + 1) \leq k \leq \lambda_1 + 1$.

Since $\lambda_1 \leq 5$, we have $6 \leq n \leq 12$ and there are finitely many possibilities for (x, n, k). In the tables below we list all possible triples (x, n, k) which satisfy none of conditions (B),(C),(D) and show that each such triple satisfies (A).

$n = 6$

$shape(x)$	k	$f_2(x)/\binom{n}{2}$	$f_k(x)/\binom{n}{k}$
$1^3 3^1$	3	$\frac{1}{5}$	$\frac{1}{10}$
$1^2 4^1$	3	$\frac{1}{15}$	0
$2^1 4^1$	3	$\frac{1}{15}$	0

$n = 7$

$shape(x)$	k	$f_2(x)/\binom{n}{2}$	$f_k(x)/\binom{n}{k}$
$1^4 3^1$	3, 4	$\frac{2}{7}$	$\frac{1}{7}$
$1^3 4^1$	3, 4	$\frac{1}{7}$	$\frac{1}{35}$
$1^2 2^1 3^1$	3, 4	$\frac{2}{21}$	$\frac{3}{35}$
$1^2 5^1$	3, 4	$\frac{1}{21}$	0
$1^1 2^1 4^1$	3, 4	$\frac{1}{21}$	$\frac{1}{35}$
$2^2 3^1$	3, 4	$\frac{2}{21}$	$\frac{1}{35}$
$2^1 5^1$	3, 4	$\frac{1}{21}$	0

$n = 8$

$shape(x)$	k	$f_2(x)/\binom{n}{2}$	$f_k(x)/\binom{n}{k}$
$1^5 3^1$	4	$\frac{5}{14}$	$\frac{1}{7}$
$1^4 4^1$	3, 5	$\frac{3}{14}$	$\frac{1}{14}$
$1^4 4^1$	4	$\frac{3}{14}$	$\frac{1}{35}$
$1^3 2^1 3^1$	4	$\frac{1}{7}$	$\frac{3}{35}$
$1^3 5^1$	3, 5	$\frac{3}{28}$	$\frac{1}{56}$
$1^3 5^1$	4	$\frac{3}{28}$	0
$1^2 2^1 4^1$	3, 5	$\frac{1}{14}$	$\frac{1}{28}$
$1^2 2^1 4^1$	4	$\frac{1}{14}$	$\frac{1}{35}$
$1^1 2^2 3^1$	4	$\frac{1}{14}$	$\frac{1}{35}$
$1^1 2^1 5^1$	3, 5	$\frac{1}{28}$	$\frac{1}{56}$
$1^1 2^1 5^1$	4	$\frac{1}{28}$	0
$2^2 4^1$	3, 5	$\frac{1}{14}$	0
$2^2 4^1$	4	$\frac{1}{14}$	$\frac{1}{35}$
$2^1 3^2$	4	$\frac{1}{14}$	0

6. ACTIONS ON 3-SETS

$n = 9$

$shape(x)$	k	$f_2(x)/\binom{n}{2}$	$f_k(x)/\binom{n}{k}$
$1^5 4^1$	4,5	$\frac{5}{18}$	$\frac{1}{21}$
$1^4 5^1$	3,6	$\frac{1}{6}$	$\frac{1}{21}$
$1^4 5^1$	4,5	$\frac{1}{6}$	$\frac{1}{126}$
$1^3 2^1 4^1$	4,5	$\frac{1}{9}$	$\frac{2}{63}$
$1^2 2^1 5^1$	3,6	$\frac{1}{18}$	$\frac{1}{42}$
$1^2 2^1 5^1$	4,5	$\frac{1}{18}$	$\frac{1}{126}$
$1^2 3^1 4^1$	4,5	$\frac{1}{36}$	$\frac{1}{42}$
$1^1 2^2 4^1$	4,5	$\frac{1}{18}$	$\frac{1}{63}$
$2^2 5^1$	3,6	$\frac{1}{18}$	0
$2^2 5^1$	4,5	$\frac{1}{18}$	$\frac{1}{126}$
$2^1 3^1 4^1$	4,5	$\frac{1}{36}$	$\frac{1}{126}$

$n = 10$

$shape(x)$	k	$f_2(x)/\binom{n}{2}$	$f_k(x)/\binom{n}{k}$
$1^6 4^1$	5	$\frac{1}{3}$	$\frac{1}{21}$
$1^5 5^1$	4,6	$\frac{2}{9}$	$\frac{1}{42}$
$1^5 5^1$	5	$\frac{2}{9}$	$\frac{1}{126}$
$1^4 2^1 4^1$	5	$\frac{7}{45}$	$\frac{2}{63}$
$1^3 2^1 5^1$	4,6	$\frac{4}{45}$	$\frac{1}{70}$
$1^3 2^1 5^1$	5	$\frac{4}{45}$	$\frac{1}{126}$
$1^3 3^1 4^1$	5	$\frac{1}{15}$	$\frac{1}{42}$
$1^2 2^2 4^1$	5	$\frac{1}{15}$	$\frac{1}{63}$
$1^2 3^1 5^1$	4,6	$\frac{1}{45}$	$\frac{1}{105}$
$1^2 3^1 5^1$	5	$\frac{1}{45}$	$\frac{1}{126}$
$1^2 4^2$	5	$\frac{1}{45}$	$\frac{1}{63}$
$1^1 2^1 3^1 4^1$	5	$\frac{1}{45}$	$\frac{1}{126}$
$1^1 2^2 5^1$	4,6	$\frac{2}{45}$	$\frac{1}{210}$
$1^1 2^2 5^1$	5	$\frac{2}{45}$	$\frac{1}{126}$
$2^3 4^1$	5	$\frac{1}{15}$	0
$2^1 3^1 5^1$	4,6	$\frac{1}{45}$	0
$2^1 3^1 5^1$	5	$\frac{1}{45}$	$\frac{1}{126}$
$2^1 4^2$	5	$\frac{1}{45}$	0

$n = 11$

$shape(x)$	k	$f_2(x)/\binom{n}{2}$	$f_k(x)/\binom{n}{k}$
$1^6 5^1$	5,6	$\frac{3}{11}$	$\frac{1}{66}$
$1^4 2^1 5^1$	5,6	$\frac{7}{55}$	$\frac{5}{462}$
$1^3 3^1 5^1$	5,6	$\frac{3}{55}$	$\frac{2}{231}$
$1^2 2^2 5^1$	5,6	$\frac{3}{55}$	$\frac{1}{154}$
$1^2 4^1 5^1$	5,6	$\frac{1}{55}$	$\frac{1}{154}$
$1^1 2^1 3^1 5^1$	5,6	$\frac{1}{55}$	$\frac{1}{231}$
$2^3 5^1$	5,6	$\frac{3}{55}$	$\frac{1}{462}$
$2^1 4^1 5^1$	5,6	$\frac{1}{55}$	$\frac{1}{462}$

$n = 12$

shape(x)	k	$f_2(x)/\binom{n}{2}$	$f_k(x)/\binom{n}{k}$
$1^7 5^1$	6	$\frac{7}{22}$	$\frac{1}{66}$
$1^5 2^1 5^1$	6	$\frac{1}{6}$	$\frac{5}{462}$
$1^4 3^1 5^1$	6	$\frac{1}{11}$	$\frac{2}{231}$
$1^3 2^2 5^1$	6	$\frac{5}{66}$	$\frac{1}{154}$
$1^3 4^1 5^1$	6	$\frac{1}{22}$	$\frac{1}{154}$
$1^2 2^1 3^1 5^1$	6	$\frac{1}{33}$	$\frac{1}{231}$
$1^2 5^2$	6	$\frac{1}{66}$	$\frac{1}{231}$
$1^1 2^3 5^1$	6	$\frac{1}{22}$	$\frac{1}{462}$
$1^1 2^1 4^1 5^1$	6	$\frac{1}{66}$	$\frac{1}{462}$
$2^2 3^1 5^1$	6	$\frac{1}{33}$	0
$2^1 5^2$	6	$\frac{1}{66}$	0

Now assume that k is either too big or too small but not both. If k is too big then $n-k$ is too small. Since $\delta_{2,k}(x) = \delta_{2,n-k}(x)$ and (x,n,k) satisfies one of conditions (B),(C),(D) if and only if $(x,n,n-k)$ satisfies the same condition, we may (and do) assume without loss of generality that k is too small, that is,

- $3 \leq k \leq \lambda_1 + 1$, and
- $k \leq n - \lambda_1 - 2$.

Since k is not too big, $(x^-, n-\lambda_1, k)$ satisfies one of (A)-(D) by inductive hypothesis. If $(x^-, n-\lambda_1, k)$ satisfies (B), so $f_2(x^-) = 0$, then $f_2(x) = 0$ and (x,n,k) satisfies (B). If $(x^-, n-\lambda_1, k)$ satisfies (C) then $k = 3$ and x has shape $1^1 2^1 3^{\frac{n-3-\lambda_1}{3}} \lambda_1^1$. If $\lambda_1 = 3$ then (x,n,k) satisfies (C). Otherwise,

$$\frac{f_3(x)}{\binom{n}{3}} = \frac{2(n-\lambda_1)}{n(n-1)(n-2)} < \frac{2}{n(n-1)} = \frac{f_2(x)}{\binom{n}{2}}$$

and (x,n,k) satisfies (A). If $(x^-, n-\lambda_1, k)$ satisfies (D) then $k = 4$ and one of the cases in the table below holds.

shape(x)	$\frac{f_2(x)}{\binom{n}{2}}$	$\frac{f_4(x)}{\binom{n}{4}}$
$1^2 3^3$	$\frac{1}{55}$	$\frac{1}{55}$
$1^2 3^2 4^1$	$\frac{1}{66}$	$\frac{1}{99}$
$1^2 3^2 5^1$	$\frac{1}{78}$	$\frac{4}{715}$

In each case $(x,n,4)$ satisfies (A). So, we assume now that $(x^-, n-\lambda_1, k)$ satisfies (A). By inductive hypothesis, we have

$$\frac{f_k(x^-)}{\binom{n}{k}} = \frac{f_k(x^-)}{\binom{n-\lambda_1}{k}} \frac{\binom{n-\lambda_1}{k}}{\binom{n}{k}}$$

$$\leq \frac{f_2(x^-)}{\binom{n-\lambda_1}{2}} \frac{\binom{n-\lambda_1}{k}}{\binom{n}{k}}$$

$$= \frac{f_2(x)}{\binom{n}{2}} \frac{\binom{n}{2}}{\binom{n-\lambda_1}{2}} \frac{\binom{n-\lambda_1}{k}}{\binom{n}{k}}$$

$$= \frac{f_2(x)}{\binom{n}{2}} \prod_{j=0}^{\lambda_1 - 1} \frac{n-k-j}{n-2-j}.$$

Now if $k < \lambda_1$ then $f_{k-\lambda_1}(x^-) = 0$. Therefore
$$\frac{f_k(x)}{\binom{n}{k}} = \frac{f_2(x)}{\binom{n}{2}} \prod_{j=0}^{\lambda_1-1} \frac{n-k-j}{n-2-j} < \frac{f_2(x)}{\binom{n}{2}}$$
and (x,n,k) satisfies condition (A).

If $k = \lambda_1$ then
$$f_{k-\lambda_1}(x^-) = f_0(x^-) = 1,$$
so
$$\frac{f_k(x)}{\binom{n}{k}} \le \frac{1}{\binom{n}{k}} + \frac{f_2(x)}{\binom{n}{2}} \prod_{j=0}^{\lambda_1-1} \frac{n-k-j}{n-2-j}.$$
Given that $3 \le k \le n-3$, we see that the right side of the last inequality is maximized when $k = 3$, so
$$\frac{f_k(x)}{\binom{n}{k}} \le \frac{1}{\binom{n}{3}} + \frac{f_2(x)}{\binom{n}{2}} \frac{n-5}{n-2}.$$
If $f_2(x) = 0$ then (x,n,k) satisfies condition (B). Otherwise,
$$\frac{f_2(x)}{\binom{n}{2}} - \frac{f_k(x)}{\binom{n}{k}} \ge \frac{f_2(x)}{\binom{n}{2}} \frac{3}{n-2} - \frac{1}{\binom{n}{3}}$$
$$\ge \frac{2}{n(n-1)} \frac{3}{n-2} - \frac{6}{n(n-1)(n-2)}$$
$$= 0$$
and (x,n,k) satisfies (A).

If $k = \lambda_1 + 1$ then
$$f_{k-\lambda_1}(x^-) = f_1(x^-) = f_1(x),$$
so
$$\frac{f_k(x)}{\binom{n}{k}} \le \frac{f_1(x)}{\binom{n}{k}} + \frac{f_2(x)}{\binom{n}{2}} \prod_{j=0}^{\lambda_1-1} \frac{n-k-j}{n-2-j}.$$
It follows that (x,n,k) satisfies condition (A) as long as
$$(6.7) \qquad k! f_1(x) \le 2(n-k-1) f_2(x) \left(\prod_{j=0}^{k-2}(n-2-j) - \prod_{j=0}^{k-2}(n-k-j) \right).$$

If $f_2(x) = 0$ then (x,n,k) satisfies (B), so we assume from now on that $f_2(x) \ge 1$. If $f_1(x) = 0$ then certainly inequality (6.7) is satisfied. Say $f_1(x) = 1$. Then inequality (6.7) is satisfied whenever
$$(6.8) \qquad k! \le 2(n-k-1) \left(\prod_{j=0}^{k-2}(n-2-j) - \prod_{j=0}^{k-2}(n-k-j) \right).$$

If $f_1(x) \ge 2$ then it follows from the fact that $f_2(x) \ge \binom{f_1(x)}{2}$ that (6.7) holds whenever
$$(6.9) \qquad k! \le (n-k-1) \left(\prod_{j=0}^{k-2}(n-2-j) - \prod_{j=0}^{k-2}(n-k-j) \right).$$

Thus it suffices to show that (6.9) holds whenever $k \in \{4,5,6\}$ and $n \geq k+3$. Substituting $k = 4,5,6$ into (6.8) gives the inequalities in the table below.

k	(6.9)
4	$(n-4)^2(n-5) \geq 4$
5	$(n-5)(n-6)(n^2 - 10n + 26) \geq 10$
6	$(n-6)(n-7)(n^3 - 18n^2 + 113n - 246) \geq 36$

Since $k \leq n-3$, all of the inequalities in the table hold and (x,n,k) satisfies (A).

We may (and do) now assume that k is neither too big nor to small, that is,

- $2 \leq k - \lambda_1 \leq k \leq n - \lambda_1 - 2$.

By inductive hypothesis, $(x^-, n - \lambda_1, k)$ satisfies one of (A)-(D), as does $(x^-, n - \lambda_1, k - \lambda_1)$. If both triples satisfy (A) then (x,n,k) satisfies (A), by fact (∗). If either triple satisfies (B) then (x,n,k) satisfies (B). Since k is neither too big too small and $\lambda_1 \geq 3$, we have $5 \leq k \leq n-5$, so $(x^-, n-\lambda_1, k)$ satisfies neither (C) nor (D) and $(x^-, n - \lambda_1, k - \lambda_1)$ does not satisfy (D)

Say $(x^-, n - \lambda_1, k - \lambda_1)$ satisfies (C). Then one of the cases in the table below holds.

λ_1	k	$f_k(x)$
3	6	$\frac{n^2 - 3n}{18}$
4	7	$\frac{n^2 - 11n + 46}{18}$
5	8	$\frac{n^2 - 13n + 58}{18}$

It follows that (x,n,k) satisfies condition (A) if and only if one of the inequalities in the table below holds.

λ_1	Inequality for (A)
3	$n^4 - 14n^3 + 51n^2 - 154n + 180 \geq 0$
4	$n^5 - 20n^4 + 155n^3 - 720n^2 + 2584n - 7160 \geq 0$
5	$n^6 - 27n^5 + 295n^4 - 1665n^3 + 3984n^2 + 6532n - 59920 \geq 0$

Each of the inequalities in the table holds for all $n \geq 12$, and since $k \leq n-5$ we see that (x,n,k) satisfies condition (A) in all cases.

It remains to show that $\delta_{2,k}(x) \geq -\frac{2}{(n-1)(n-2)}$ if (x,n,k) satisfies one of conditions (B),(C),(D). Say $f_2(x) = 0$. Then

$$(c_1(x), c_2(x)) \in \{(0,0), (1,0)\}.$$

We may (and do) assume that $k \leq \frac{n}{2}$. As above, let x have cycle shape $(\lambda_1, \ldots, \lambda_r)$. Define

$$\mathcal{A}(x) := \left\{ I \subseteq [r] : \sum_{i \in I} \lambda_i = k \right\}.$$

Then $f_k(x) = |\mathcal{A}(x)|$. Note that $\mathcal{A}(x)$ is an antichain, that is, no element of $\mathcal{A}(x)$ is contained in another element. It follows from a general version of Sperner's theorem (see [**Bo**, Chapter 3, Theorem 2]) that if each element of $\mathcal{A}(x)$ has size at most m and $m \leq \frac{r}{2}$ then $|\mathcal{A}(x)| \leq \binom{r}{m}$.

If $(c_1(x), c_2(x)) = (0,0)$ then $r \leq \lfloor \frac{n}{3} \rfloor$ and for each $I \in \mathcal{A}(x)$ we have $|I| \leq \lfloor \frac{k}{3} \rfloor$. Therefore, if $\lfloor \frac{k}{3} \rfloor \leq \frac{r}{2}$ then

$$f_k(x) \leq \binom{\lfloor \frac{n}{3} \rfloor}{\lfloor \frac{k}{3} \rfloor}.$$

If $\lfloor \frac{k}{3} \rfloor > \frac{r}{2}$ then by the usual Sperner theorem we have

$$f_k(x) \leq \binom{r}{\lfloor \frac{r}{2} \rfloor}$$
$$< \binom{2\lfloor \frac{k}{3} \rfloor}{\lfloor \frac{k}{3} \rfloor}$$
$$\leq \binom{\lfloor \frac{n}{3} \rfloor}{\lfloor \frac{k}{3} \rfloor},$$

the last inequality holding under our assumption that $k \leq \frac{n}{2}$.

If $(c_1(x), c_2(x)) = (1, 0)$ then $\lambda_r = 1$. Define

$$\mathcal{A}'(x) := \left\{ I \subseteq [r-1] : \sum_{i \in I} \lambda_i \in \{k-1, k\} \right\}.$$

Then $\mathcal{A}'(x)$ is an antichain and $f_k(x) = |\mathcal{A}'(x)|$. Also, $r - 1 \leq \lfloor \frac{n-1}{3} \rfloor \leq \lfloor \frac{n}{3} \rfloor$ and every element of $\mathcal{A}'$ has size at most $\lfloor \frac{k}{3} \rfloor$. It now follows from the same arguments we used in the case $(c_1(x), c_2(x)) = (0, 0)$ that

$$f_k(x) \leq \binom{\lfloor \frac{n}{3} \rfloor}{\lfloor \frac{k}{3} \rfloor}.$$

For nonnegative integers n, k, define

$$\phi(n, k) := \frac{\binom{\lfloor \frac{n}{3} \rfloor}{\lfloor \frac{k}{3} \rfloor}}{\binom{n}{k}}.$$

It now suffices to show that the inequality

(6.10) $$\phi(n, k) \leq \frac{2}{(n-1)(n-2)}$$

holds whenever $3 \leq k \leq \frac{n}{2}$. Now

$$\phi(n, 3) = \frac{\lfloor \frac{n}{3} \rfloor}{\binom{n}{3}} \leq \frac{2}{(n-1)(n-2)},$$

$$\phi(n, 4) = \frac{4}{n-3} \phi(n, 3),$$

and

$$\phi(n, 5) = \frac{20}{(n-3)(n-4)} \phi(n, 3).$$

Under the assumption that $k \leq n - 3$, it follows that inequality (6.10) holds whenever $k \in \{3, 4, 5\}$, so it now suffices to show that $\phi(n, k+3) \leq \phi(n, k)$ whenever $3 \leq k \leq \frac{n}{2} - 3$. Define

$$\psi(n, k) := \frac{(\frac{n}{3} - \lfloor \frac{k}{3} \rfloor)(k+3)(k+2)(k+1)}{(\lfloor \frac{k}{3} \rfloor + 1)(n-k)(n-k-1)(n-k-2)}.$$

Then

$$\frac{\phi(n,k+3)}{\phi(n,k)} = \frac{\binom{\lfloor \frac{n}{3} \rfloor}{\lfloor \frac{k}{3} \rfloor+1}\binom{n}{k}}{\binom{\lfloor \frac{n}{3} \rfloor}{\lfloor \frac{k}{3} \rfloor}\binom{n}{k+3}}$$

$$= \frac{(\lfloor \frac{n}{3} \rfloor - \lfloor \frac{k}{3} \rfloor)(k+3)(k+2)(k+1)}{(\lfloor \frac{k}{3} \rfloor+1)(n-k)(n-k-1)(n-k-2)}$$

$$\leq \psi(n,k).$$

Now

$$\psi(n,k) = \begin{cases} \frac{(k+2)(k+1)}{(n-k-1)(n-k-2)}, & k \equiv 0 \bmod 3, \\ \frac{(n-k+1)(k+3)(k+1)}{(n-k)(n-k-1)(n-k-2)}, & k \equiv 1 \bmod 3, \\ \frac{(n-k+2)(k+3)(k+2)}{(n-k)(n-k-1)(n-k-2)}, & k \equiv 2 \bmod 3. \end{cases}$$

It is now straightforward to verify that under the assumption $3 \leq k \leq \frac{n}{2} - 3$ we have $\psi(n,k) < 1$. For instance, one can show that for fixed n, ψ increases as k increases and that $\psi(n, \frac{n}{2} - 3) < 1$. Therefore, if condition (B) is satisfied then $\delta_{2,k}(x) \geq -\frac{2}{(n-1)(n-2)}$ as claimed.

If x has shape $1^1 2^1 3^{(n-3)/3}$ then

$$\delta_{2,3}(x) = \frac{1}{\binom{n}{2}} - \frac{\frac{n}{3}}{\binom{n}{3}}$$

$$= -\frac{4}{n(n-1)(n-2)}$$

$$> -\frac{2}{(n-1)(n-2)},$$

and if x has shape $1^2 3^2$ then

$$\delta_{2,4}(x) = \frac{1}{28} - \frac{4}{70}$$

$$= -\frac{3}{140}$$

$$> -\frac{2}{42},$$

so $\delta_{2,k}(x) > -\frac{2}{(n-1)(n-2)}$ if one of conditions (C),(D) is satisfied. This completes the proof of Lemma 6.0.52. □

Note that if x has shape $3^{n/3}$ then $\delta_{2,3}(x) = -\frac{2}{(n-1)(n-2)}$, so the bound given in the lemma the best one possible.

LEMMA 6.0.53. *For any $n \geq 5$ and any $x \in S_n$, one of the following conditions holds.*

(1) *We have $\varepsilon_{2,3}(x) \geq 0$.*
(2) *We have $o_1(x) \leq \frac{n+3}{3}$ and*

$$\varepsilon_{2,3}(x) \geq -\frac{4}{3(n-1)(n-2)}.$$

PROOF. Using Burnside's lemma and Lemma 6.0.52, we see that $\varepsilon_{2,3}(x) \geq 0$ unless some power of x fixes no 2-set or has shape $1^1 2^1 3^{(n-3)/3}$. In the second case, exactly two powers of x have shape $1^1 2^1 3^{(n-3)/3}$. Moreover, we must have $f_1(x) = c_2(x) = 1$, and it follows that $f_2(x^j) > 0$ for all j. It now follows from Lemma 6.0.52 and Burnside's lemma that

$$\begin{aligned}
\varepsilon_{2,3}(x) &\geq \frac{1}{|x|}\left(2\frac{-4}{n(n-1)(n-2)}\right) \\
&\geq -\frac{4}{3n(n-1)(n-2)} \\
&> -\frac{4}{3(n-1)(n-2)}.
\end{aligned}$$

In addition, every cycle of x besides the fixed point and the two cycle has length divisible by three. It follows that $o_1(x) \leq \frac{n+3}{3}$.

Now assume that some power of x fixes no two set. Then x fixes no two set, so $c_1(x) \leq 1$ and $c_2(x) = 0$. This gives $o_1(x) \leq \frac{n+2}{3}$. Let P be the set of powers of x which fix no two set. Noting that no power of x has shape $1^1 2^1 3^{(n-3)/3}$ and using Burnside's lemma and Lemma 6.0.52, we get

$$\begin{aligned}
\varepsilon_{2,3}(x) &\geq -\frac{1}{|x|}\sum_{x^j \in P}\frac{c_3(x^j)}{\binom{n}{3}} \\
&\geq -\frac{1}{|x|}\sum_{j=1}^{|x|}\frac{c_3(x^j)}{\binom{n}{3}}.
\end{aligned}$$

Now any three cycle in any power of x arises as a cycle in a power of a unique cycle σ of x such that $m = |supp(\sigma)|$ is divisible by three. Moreover, for any such cycle σ, exactly two powers of σ, namely, $\sigma^{m/3}$ and $\sigma^{2m/3}$, have 3-cycles in their cycle decomposition (and each has $\frac{m}{3}$ such 3-cycles). Each 3-cycle obtained from σ in this way will appear in $\frac{|x|}{m}$ powers of x. We now have

$$\begin{aligned}
\varepsilon_{2,3}(x) &\geq -\frac{1}{|x|}\frac{1}{\binom{n}{3}}\sum_{3|m}\frac{|x|}{m}2c_m(x)\frac{m}{3} \\
&= -\frac{1}{\binom{n}{3}}\sum_{3|m}\frac{2c_m(x)}{3} \\
&\geq -\frac{4}{3(n-1)(n-2)}.
\end{aligned}$$

$\square$

Note that the bound $\varepsilon_{2,3}(x) = -\frac{4}{3(n-1)(n-2)}$ is achieved when x has shape $3^{n/3}$ and therefore cannot be improved. Before continuing with our proof of Theorem 1.1.2, we record the following simpler consequence of Lemma 6.0.52, which follows from the lemma and Burnside's lemma and will be used later in the paper.

LEMMA 6.0.54. *Let $x \in S_n$ with $n \neq 8$, and assume that $3 \leq k \leq n-3$. Then*
(1) *If $f_2(x) > 0$ then $\varepsilon_{2,k}(x) \geq 0$.*
(2) *In any case, we have $\varepsilon_{2,k}(x) > -\frac{2}{(n-1)(n-2)}$.*

LEMMA 6.0.55. *Theorem 4.0.31 is true if we assume that condition (1) of Theorem 4.0.30 holds.*

PROOF. Using Lemma 2.0.6, we get

$$g_3(E) - g_2(E) = \frac{n-5}{3}(g_2(E) - 1) + \frac{n(n-1)(n-2)}{12}\sum_{x \in E}\varepsilon_{2,3}(x).$$

Set

$$F := \{x \in E : \varepsilon_{2,3}(x) < 0\}.$$

Note that since condition (1) of Theorem 4.0.30 holds, there is some constant c such that $g_2(E) > \max\{2, cn\}$, so $g_2(E) - 1 \geq 2$. Thus if $|F| \leq 3$ then Lemma 6.0.53 gives

$$\begin{aligned}g_3(E) - g_2(E) &\geq \frac{1}{3}\max\{2n - 10, cn^2 - (1+5c)n + 5\} - 3\frac{n}{9}\\&= \frac{1}{3}\max\{n - 10, cn^2 - (2+5c)n + 5\}.\end{aligned}$$

Also, for each $x \in F$, Lemma 6.0.53 gives

$$i_1(x) \geq \frac{2n-3}{3}.$$

Thus, by the Riemann-Hurwitz formula, we have

$$g_1(E) \geq 1 - n + |F|\frac{2n-3}{6}.$$

Since condition (1) of Theorem 4.0.30 holds, we have

$$g_2(E) \geq 4 - n + |F|\frac{2n-3}{6}.$$

Now, again applying Lemma 6.0.53, we get

$$\begin{aligned}g_3(E) - g_2(E) &\geq |F|\left(\frac{(n-5)(2n-3)}{18} - \frac{n}{9}\right) - \frac{(n-5)(n-3)}{3}\\&= |F|\frac{2n^2 - 15n + 15}{18} - \frac{n^2 - 8n + 15}{3}.\end{aligned}$$

Thus if $|F| \geq 4$ we get

$$g_3(E) - g_2(E) \geq \frac{n^2 - 6n - 15}{9} > \frac{n-10}{3}.$$

□

LEMMA 6.0.56. *Theorem 4.0.31 is true if we assume that condition (2) of Theorem 4.0.30 holds.*

PROOF. Direct calculation shows that if $x \in S_n$ has shape $1^a 2^b$ then

$$i_3(x) = \frac{1}{2}\left[\binom{n}{3} - \binom{a}{3} - ab\right].$$

Direct calculation now shows that if condition (2a) of Theorem 4.0.30 holds then

$$g_3(E) = \frac{(n-3)(n-5)}{4}$$

while if condition (2b) holds then
$$g_3(E) = \frac{(n-4)^2}{4}.$$

□

CHAPTER 7

Nine or more branch points - the proof of Theorem 4.0.34

In this chapter we prove Theorem 4.0.34. So, assume E, n satisfy the conditions of the theorem. As in the proof of Theorem 4.0.30, let
$$A := \{x \in E : f_1(x) \leq 4\}.$$
Assume first that $|A| \leq 3$. If $g_1(E) > 0$ then Equation (5.1) of Lemma 5.0.43 and Lemma 5.0.40(1,2,3,8a,9) give
$$g_2(E) - g_1(E) \geq \frac{n-3}{2}(g_1(E) - 1) + \frac{(r-3)(n-3)}{4} - \frac{3n}{8} \geq g_1(E) - 1 + \frac{9(n-4)}{8}.$$
This gives $g_2(E) - g_1(E) > 2$ unless $n = 5$ and $g_1(E) = 1$. Also, Equation (5.2) of Lemma 5.0.43 and Lemma 5.0.40(1,2,3,8a,9) give
$$g_2(E) - \frac{r}{3} \geq 1 - \frac{r}{3} + \frac{(r-3)(n-3)}{4} - \frac{3n}{8}.$$
It follows that $g_2(E) - \frac{r}{3} > 0$ whenever $n > 5$. If $g_1(E) = 0$ then Lemmas 5.0.43, 5.0.40(1,2,3,8a,9) and 5.0.42 give
$$g_2(E) - g_1(E) \geq -\frac{n-3}{2} + \frac{(r-3)(n-3)}{4} - \frac{n}{8} \geq \frac{7n-24}{8}$$
and
$$g_2(E) - \frac{r}{3} \geq 1 - \frac{r}{3} - \frac{n-1}{2} + \frac{(r-3)(n-3)}{4} - \frac{n}{8} \geq \frac{7n}{8} - 6.$$
Thus $g_2(E) - g_1(E) > 2$ if $n > 5$ and $g_2(E) - \frac{r}{3} > 0$ if $n > 6$.

Now assume that $|A| \geq 4$ and $n > 5$. We use Lemma 5.0.44, which is still true under the conditions of Theorem 4.0.34. We have
$$\Psi(n, 9, 4) = \frac{4n - 15}{2} > 2$$
and
$$\Gamma(n, 9, 4) = 2n - 11.$$
Thus $g_2(E) - g_1(E) > 2$ and $g_2(E) - \frac{r}{3} > 0$ when $n > 5$.

It is now sufficient to show that $g_2(E) - g_1(E) > 2$ if $n = 5$ and that $g_2(E) - \frac{r}{3} > 0$ if $n \leq 6$. By Lemma 2.0.19, we have $i_2(x) \geq n - 2$ for all $x \in E$ in any case. The Riemann-Hurwitz formula then gives
$$g_2(E) \geq \begin{cases} \frac{3r}{2} - 9 & n = 5, \\ 2r - 14 & n = 6. \end{cases}$$
The desired conclusions follow immediately.

CHAPTER 8

Actions on cosets of some 2-homogeneous and 3-homogeneous groups

In this chapter we examine the numbers $g(G, H, E)$ when $G \in \{A_n, S_n\}$ and $|E| \geq 5$ as usual, and H is a 3-homogeneous (maximal) subgroup of G. In addition, we examine the case where H is 2-homogeneous for small n. We will prove Theorem 4.0.33 and Proposition 4.0.35, along with similar results in the cases where $|E| \leq 4$ but E contains an n-cycle.

We use the following notation.

- For $x \in G$, $f_H(x)$ is the number of cosets of H fixed by x in the usual translation action, $o_H(x)$ is the number of orbits of $\langle x \rangle$ on the coset space $H \setminus G$, $i_H(x) := [G : H] - o_H(x)$ is the index of x for the given action, and $Cl_G(x)$ is the conjugacy class of x in G.

With this notation, the Riemann-Hurwitz formula says that

(8.1) $$2(g(G, H, E) + [G : H] - 1) = \sum_{x \in E} i_H(x).$$

By Burnside's lemma, we have

(8.2) $$o_H(x) = \frac{1}{|x|} \sum_{j=1}^{|x|} f_H(x^j),$$

and in order to get lower bounds on $g(G, H, E)$ we will find upper bounds on $f_H(x)$ for appropriate $x \in S_n$. Straightforward calculation gives

(8.3) $$f_H(x) = [G : H] \frac{|Cl_G(x) \cap H|}{|Cl_G(x)|}.$$

We define, for any $H \leq G$,

$$m(G, H) := \max_{g \in G \setminus \{1\}} \frac{|Cl_G(x) \cap H|}{|Cl_G(x)|}.$$

Using (8.1, 8.2, 8.3) and direct calculation, we get the following result.

PROPOSITION 8.0.57. *For any monodromy data (G, H, E), we have*

$$g(G, H, E) \geq 1 + \frac{[G : H]}{2} \left(\left(|E| - \sum_{x \in E} \frac{1}{|x|} \right) [1 - m(G, H)] - 2 \right)$$

As we shall see, Proposition 8.0.57 and upper bounds on $m(G, H)$ for most H under consideration in this chapter will give the results we desire in many cases, but additional work will be needed when n or $|E|$ is small. We now proceed to find upper bounds on $m(G, H)$, starting by finding lower bounds on $|Cl_{S_n}(x)|$ for certain x.

LEMMA 8.0.58. *Let $x \in S_n$ have cycle shape $1^a p^{(n-a)/p}$. Then*

$$|Cl_{S_n}(x)| \geq \frac{e}{2} \left(\frac{n-a}{e}\right)^{(n-a)/2} \binom{n}{a}.$$

PROOF. Let $x^* \in S_{n-a}$ have shape $p^{(n-a)/p}$. A simple calculation gives

$$|Cl_{S_n}(x)| = \binom{n}{a} |Cl_{S_{n-a}}(x^*)|.$$

It therefore suffices to show that for integers m, b with $1 \leq b \leq \frac{m}{2}$, we have

(8.4) $$\phi(m, b) := \frac{m!}{b! \left(\frac{m}{b}\right)^b} \left(\frac{e}{m}\right)^{m/2} \geq \frac{e}{2}.$$

For fixed b and all $m \geq 2b$, we have

$$\frac{\phi(m+1, b)}{\phi(m, b)} = \sqrt{e(m+1)} \left(\frac{m}{m+1}\right)^{b+\frac{m}{2}} \geq \sqrt{e(m+1)} \left(\frac{m}{m+1}\right)^m$$
$$> \sqrt{\frac{m+1}{e}} > 1,$$

so it suffices to show that (8.4) holds when $m = 2b$. When $b = 1$ and $m = 2$, (8.4) holds with equality. Assume $b > 1$ and set

$$\rho(b) := \frac{2\phi(2b, b)}{e},$$

and

$$\alpha(b) := \frac{\rho(b+1)}{\rho(b)} = e\frac{2b+1}{2b+2} \left(\frac{b}{b+1}\right)^b.$$

Then (considering α as a function on $[1.9, \infty)$), we have

$$\alpha'(b) = e\frac{b^b}{(b+1)^{(b+1)}} \left[1 - \left(b + \frac{1}{2}\right) \ln\left(1 + \frac{1}{b}\right)\right] < 0.$$

(To show that the last inequality holds, one can show that $\alpha'(1.9) < 0$ and $(b + \frac{1}{2}) \ln(1 + \frac{1}{b})$ decreases to one as b increases to ∞.) Since $\alpha(2) = \frac{10e}{27} > 1$ and $\lim_{b \to \infty} \alpha(b) = 1$, we have $\rho(b+1) > \rho(b)$ for all $b \geq 2$, and since $\rho(2) = \frac{6e}{16} > 1$, we are done. □

COROLLARY 8.0.59. *Say $n > 5$ and $x \in S_n$ has prime order and fixes at most $\frac{n}{2}$ points. Then*

$$|Cl_{S_n}(x)| \geq \frac{e}{2} \left(\frac{2n}{e}\right)^{n/4}.$$

PROOF. If x fixes no point then Lemma 8.0.58 gives

$$|Cl_{S_n}(x)| \geq \frac{e}{2} \left(\frac{n}{e}\right)^{n/2} > \frac{e}{2} \left(\frac{2n}{e}\right)^{n/4},$$

the second inequality holding under our assumption that $n > 5$. Now assume that x fixes at least one point. For $a \in [1, \frac{n}{2}]$, set

$$\mu(a) := \frac{e}{2}\left(\frac{n-a}{e}\right)^{(n-a)/2}\left(\frac{n}{a}\right)^a.$$

If x fixes a points then $|Cl_{S_n}(x)| \geq \mu(a)$ by Lemma 8.0.58. It is straightforward to show that $\mu' < 0$ on the given domain, so μ is minimized at $\frac{n}{2}$. Since

$$\mu\left(\frac{n}{2}\right) = \frac{e}{2}\left(\frac{2n}{e}\right)^{n/4},$$

we are done. $\square$

Now let us examine 3-homogeneous subgroups of S_n. The following result is easily obtained from information in [**Ca**] and [**Ka**].

LEMMA 8.0.60. *Let $H \leq S_n$ be 3-homogeneous with $A_n \not\leq H$. Then one of the following conditions holds.*

(1) $L_2(q) \leq H \leq P\Gamma L_2(q)$ and $n = q + 1$.
(2) $H \leq AGL_k(2)$ and $n = 2^k$.
(3) H is a Mathieu group M_n.
(4) $H = M_{11}$ (acting on cosets of $L_2(11)$) and $n = 12$.
(5) $H \leq AGL_1(q)$ with $q = n \in \{8, 32\}$.

Thus all but finitely many 3-homogeneous groups (other than A_n or S_n) are either affine over $\mathbb{F}_2$ or have socle $L_2(q)$. The following result is a special case of the a result in [**GuMa**], but we include its short proof.

LEMMA 8.0.61. *If $H = AGL_k(2)$ acting on $\mathbb{F}_2^k$ or $H = P\Gamma L_2(q)$ acting on 1-spaces from $\mathbb{F}_q^2$ with $q \geq 7$ then every nonidentity element of H fixes at most half of the points in the given action.*

PROOF. Say $x \in AGL_k(2)$ maps each $w \in \mathbb{F}_2^k$ to $wA + v$ with $A \in GL_k(2)$ and $v \in \mathbb{F}_2^k$. Then x fixes w if an only if $w(A + I) = v$. Thus x is fixed-point-free if v is not in the image of $A + I$, and if v lies in this image then the set of fixed points of x is a coset of the kernel of $A + I$. This kernel has size at most 2^{k-1} unless $A = I$, in which case either $x = 1$ or x is fixed-point-free.

If $q = p^a$ with p prime then $P\Gamma L_2(q)$ is a semidirect product with kernel $PGL_2(q)$ and (possibly trivial) complement $F \cong Aut(\mathbb{F}_q)$. Note that $PGL_2(q)$ is 3-transitive in the action under consideration. Thus if $x \in P\Gamma L_2(q)$ fixes at least three points, we may assume these fixed points include the 1-spaces generated by $[1, 0]$, $[0, 1]$ and $[1, 1]$. It follows that $x \in F$ and the 1-spaces fixed by x are those generated by $[0, 1]$ and $[1, \alpha]$, where α is fixed by the element of $Aut(\mathbb{F}_q)$ corresponding to x. Now any nontrivial automorphism of $\mathbb{F}_q$ fixes at most $\sqrt{q}$ elements, and since $q \geq 7$ we have $1 + \sqrt{q} < \frac{1+q}{2}$. $\square$

COROLLARY 8.0.62. *Let $G \in \{A_n, S_n\}$, let $H < G$ be 3-homogeneous with either $n = 2^k$ ($k \geq 5$) and $H \leq AGL_k(2)$, or $n = q + 1$ ($q \geq 19$) and $L_2(q) \leq H \leq P\Gamma L_2(q)$. Let $E = (x_1, \ldots, x_r)$ be an r-tuple of elements of $G \setminus \{1\}$ such that $\langle E \rangle = G$ and $\prod_{i=1}^r x_i = 1$. There exists some constant c (independent of (G, H, E)) such that*

(1) $m(G, H) < \frac{2}{25}$,

(2) if $x \in S_n$ with $|x| \leq 6$ then
$$\frac{|Cl_G(y) \cap H|}{|Cl_G(y)|} < \frac{1}{25}$$
for all $y \in \langle x \rangle \setminus \{1\}$,

(3) if $x \in S_n$ is an n-cycle then
$$\frac{|Cl_G(y) \cap H|}{|Cl_G(y)|} < \frac{1}{10,000}$$
for all $y \in \langle x \rangle \setminus \{1\}$, and

(4) if $r \geq 5$ or x_r is an n-cycle then
$$g(G, H, E) > \max\{2, \frac{r}{3}, c[G:H]\}.$$

PROOF. We show first that the first three claims together imply the fourth. Note that if $n = 2^k$ and $H \leq AGL_k(2)$ then

(8.5) $$|H| < n^{(1+\log_2(n))},$$

while if $n = q + 1$ and $H \leq P\Gamma L_2(q)$ then

(8.6) $$|H| < n^3 \log_2(n).$$

Under our assumptions, these inequalities give
$$[G:H] > \frac{20!}{2\log_2(20) \cdot 20^3}.$$

By Proposition 8.0.57 and the first claim, we have
$$\begin{aligned} g(G, H, E) &\geq 1 + \frac{[G:H]}{2}\left(\left(r - \frac{r}{2}\right)\left(1 - \frac{2}{25}\right) - 2\right) \\ &= 1 + \frac{[G:H](23r - 100)}{100}, \end{aligned}$$

and it follows that the fourth claim holds when $r \geq 5$. If $r = 4$ and E contains an n-cycle, using our assumption that $n \geq 20$ we get
$$\begin{aligned} g(G, H, E) &\geq 1 + \frac{[G:H]}{2}\left(\left(\frac{5}{2} - \frac{1}{n}\right)\left(1 - \frac{2}{25}\right) - 2\right) \\ &\geq 1 + \frac{127[G:H]}{1,000}, \end{aligned}$$

and again the fourth claim holds. Say $r = 3$ and E contains an n-cycle. If E contains at most one element of order less than seven, then we have
$$\begin{aligned} g(G, H, E) &\geq 1 + \frac{[G:H]}{2}\left(\left(3 - \frac{1}{2} - \frac{1}{7} - \frac{1}{n}\right)\left(1 - \frac{2}{25}\right) - 2\right) \\ &\geq 1 + \frac{429[G:H]}{7,000}. \end{aligned}$$

If E contains one n cycle and two elements x_1, x_2 of order at most six, we employ the second and third claims. Arguing as we did to prove Proposition 8.0.57, and

8. 2-HOMOGENOUS AND 3-HOMOGENEOUS GROUPS

noting that E cannot contain two involutions, we see that

$$\begin{aligned} g(G,H,E) &\geq 1 + \frac{[G:H]}{2}\left(\frac{7}{6}\left(1-\frac{1}{25}\right) + \left(1-\frac{1}{n}\right)\left(1-\frac{1}{10,000}\right) - 2\right) \\ &\geq 1 + \frac{41,943[G:H]}{1,200,000}. \end{aligned}$$

The fourth claim holds in all possible cases.

Now we prove the first three claims. As is well known and easy to prove, for any $x \in S_n$, we have

$$|Cl_G(x)| = \begin{cases} |Cl_{S_n}(x)|/2 & \text{if } G = A_n \text{ and } x \text{ is the product of disjoint} \\ & \text{cycles of pairwise distinct odd lengths,} \\ |Cl_{S_n}(x)| & \text{otherwise} \end{cases}$$

Certainly

$$\frac{|Cl_G(x) \cap H|}{|Cl_G(x)|} < \frac{|H|}{|Cl_G(x)|}$$

for all $x \in S_n$. If $n > 20$ and $|x| < 7$ then x is not the product of disjoint cycles of pairwise distinct odd lengths, and the first two claims now follow from inequalities (8.5) and (8.6), Lemma 8.0.61, Corollary 8.0.59 and the fact that the centralizer of $x \in S_n$ also centralizes every power of x (which allows us to apply Corollary 8.0.59 without knowing that $|x|$ is prime) by straightforward calculations. Finally, if $x \in S_n$ is an n-cycle then every nontrivial power of x is homocyclic and fixed-point-free. We can prove the third claim as we proved the first two, replacing Corollary 8.0.59 with Lemma 8.0.58 (using $a = 0$). □

To complete the proof of Theorem 4.0.33, we must examine the finitely many remaining pairs (G, H) such that $G \in \{A_n, S_n\}$ and $H \neq A_n$ is a 3-homogeneous maximal subgroup of G. All such pairs (G, H) are described in the table below. (Note that there are embeddings $A\Gamma L_1(2^k) \hookrightarrow AGL_k(2) \hookrightarrow A_{2^k}$ and that M_{11}

acting on cosets of $L_2(11)$ is contained in M_{12}.)

G	H	$m(G,H)$
A_6	$L_2(5)$	$1/2$
S_6	$PGL_2(5)$	$2/3$
A_8 or S_8	$L_2(7)$ or $PGL_2(7)$	$1/5$
A_8	$AGL_3(2)$	$7/15$
A_9	$P\Gamma L_2(8)$	$1/15$
A_{10}	M_{10}	$1/105$
S_{10}	$Aut(A_6)$	$4/105$
A_{11}	M_{11}	$1/105$
A_{12}	M_{12}	$4/105$
S_{12}	$PGL_2(11)$	$1/189$
A_{14}	$L_2(13)$	$1/10,395$
S_{14}	$PGL_2(13)$	$2/3,465$
A_{16}	$AGL_4(2)$	$43/135,135$
A_{17}	$P\Gamma L_2(16)$	$1/135,135$
A_{18}	$L_2(17)$	$1/2,027,025$
S_{18}	$PGL_2(17)$	$8/2,027,025$
A_{22}	M_{22}	$1/130,945,815$
S_{22}	$Aut(M_{22})$	$2/19,840,275$
A_{23}	M_{23}	$1/130,945,815$
A_{24}	M_{24}	$2/19,840,275$

Theorem 4.0.33 now follows from direct calculations. (Note that if (G,H) is one of $(A_8, L_2(7))$ or $(S_8, PGL_2(7))$ and $x \in G$, we have

$$\frac{|Cl_G(x) \cap H|}{|Cl_G(x)|} \leq \frac{1}{15}$$

unless x has cycle shape 2^4.)

The next result will be used in the proof of Theorem 1.2.1.

THEOREM 8.0.63. *Let $G \in \{S_n, A_n\}$ with $n \geq 7$, let $H < G$ be a maximal 3-homogeneous subgroup with $H \neq A_n$, let $r \in \{3,4\}$, let $E = (x_1, \ldots, x_r)$ be an n-tuple of elements of $G \setminus \{1\}$ such that $\langle E \rangle = G$, $\prod_{i=1}^r x_i = 1$ and x_r is an n-cycle. Then $g(G, H, E) > 2$.*

PROOF. By Corollary 8.0.62, we need only examine the pairs (G, H) listed in the table just above. Moreover, those cases where $G = A_n$ with n even can be ignored, since in those cases E cannot contain an n-cycle.

In each case where $n > 12$, we have $m(G, H) < \frac{1}{1,000}$ and $[G:H] \geq 11!$. We also have

$$|E| - \sum_{x \in E} \frac{1}{x} \geq \begin{cases} \frac{5}{2} - \frac{1}{n}, & r = 4, \\ \frac{13}{6} - \frac{1}{n}, & r = 3. \end{cases}$$

Using Proposition 8.0.57, we see that under our assumptions, we have

$$g(G, H, E) \geq 1 + \frac{[G:H]}{2}\left(\left(\frac{13}{6} - \frac{1}{n}\right)\left(1 - \frac{1}{1,000}\right) - 2\right)$$
$$\geq 1 + 11!\tfrac{39}{960}.$$

A similar argument gives

$$g(G,H,E) \geq \begin{cases} 1 + 9!\frac{41}{567} & (G,H) = (S_{12}, PGL_2(11)), \\ 1 + 7!\frac{194}{3,465} & (G,H) = (A_{11}, M_{11}). \end{cases}$$

We are left with $(G,H) \in \{(S_{10}, \operatorname{Aut}(A_6)), (A_9, \operatorname{Aut}(L_2(8))), (S_8, PGL_2(7))\}$. In the first case, our claim follows from the following facts, which are obtained by direct calculation:

- If $z \in S_{10}$ is a 10-cycle then $i_H(z) = 2,256$.
- For all $x \in S_{10}$, we have $i_H(x) \geq 1,212$.
- If $x \in S_{10}$ and $|x| > 2$ then $i_H(x) \geq 1,674$.

In the remaining cases, there is more work to do.

In the second case, we calculate $i_H(x)$ for $x \in A_9$ as follows.

$\operatorname{Shape}(x)$	$1^5 2^2$	$1^1 2^4$	$1^6 3^1$	$1^3 3^2$	3^3	$1^3 2^1 4^1$	$1^1 4^2$	$1^4 5^1$
$i_H(x)$	60	56	80	76	78	90	88	96

$\operatorname{Shape}(x)$	$1^2 2^2 3^1$	$1^1 2^1 6^1$	$1^2 7^1$	9^1	$2^2 5^1$	$2^1 3^1 4^1$	$1^1 3^1 5^1$
$i_H(x)$	100	96	102	104	108	110	112

We now see that if $|E| = 4$ and E contains an n-cycle then $g(G,H,E) \geq 17$. Inspection shows that if $|E| = 3$ and E contains an n-cycle x_3 then $g(G,H,E) > 2$ unless the pair of cycle shapes of elements of $E \setminus \{x_3\}$ is one of $(1^5 2^2, 1^3 3^2)$, $(1^5 2^2, 3^3)$, $(1^1 2^4, 1^6 3^1)$ or $(1^1 2^4, 3^3)$. For the first and third pairs, we have $g_1(E) = -1$ and for the second pair we have $g_2(E) = -1$. The fourth pair cannot be eliminated in this manner, but computations using MAGMA and GAP show that no triple of elements of shapes $1^1 2^4$, 3^3 and 9^1 can generate A_9 and have trivial product.

In the third case, we calculate $i_H(x)$ for $x \in G = S_8$ as follows.

$\operatorname{Shape}(x)$	$1^6 2^1$	$1^4 2^2$	$1^2 2^3$	2^4	$1^5 3^1$	$1^2 3^2$	$1^4 4^1$
$i_H(x)$	60	60	56	38	80	76	90

$\operatorname{Shape}(x)$	$2^2 4^1$	4^2	$1^3 5^1$	$1^2 6^1$	$1^1 2^2 3^1$	$2^1 3^3$	$1^2 2^1 4^1$
$i_H(x)$	96	100	98	90	82	80	90

$\operatorname{Shape}(x)$	$2^1 6^1$	$1^1 7^1$	8^1	$1^1 2^1 5^1$	$1^1 3^1 4^1$	$3^1 5^1$
$i_H(x)$	94	102	100	108	110	112

If E has size four and contains an 8-cycle x_4 then E contains either one other element not in A_8 or three such elements. In the second case, we get $i_H(x) \geq 56$ for all $x \in E \setminus \{x_4\}$, and $g(G,H,E) \geq 15$. In the first case, either $\sum_{x \in E \cap A_8} i_H(x) = 76$ or $\sum_{x \in E \cap A_8} i_H(x) \geq 98$. Direct inspection now gives $g(G,H,E) \geq 7$. If E has size three and contains an 8-cycle x_3 then exactly one of the remaining elements x_1, x_2 of E lies in A_8. Now inspection shows that if $g(G,H,E) \leq 2$ then the pair shapes of the elements of $E \setminus \{x_3\}$ is one of $(1^6 2^1, 1^5 3^1)$, $(1^6 2^1, 4^2)$, $(1^2 2^3, 4^2)$, $(2^4, 1^3 2^1 3^1)$ or $(2^4, 8)$. In the first case, we cannot have $x_1 x_2 x_3 = 1$. In the second case, we cannot have $x_1 x_2 x_3 = 1$ and $\langle E \rangle$ primitive. In the third case, we calculate $g_3(E) = 3$ and $g_4(E) = 2$. In the fourth case, we calculate $g_2(E) = -1$. In the fifth and last case we calculate $g_3(E) = 8$ and $g_4(E) = 7$. Thus we find no examples here and our proof is complete. □

Finally, we prove Theorem 4.0.35. Any 2-homogeneous subgroup of S_n is primitive. All primitive permutation groups of degree at most 20 are known (see for example [**DiMo**, Appendix B]). Examining a list of such groups, we see that the maximal 2-homogeneous H which are not 3-homogeneous are those listed in the table below.

G	H	$m(G,H)$
A_7	$L_2(7)$	$1/5$
A_9	$ASL_2(3)$	$1/40$
S_9	$AGL_2(3)$	$1/35$
A_{13}	$L_3(3)$	$1/1,155$
A_{15}	$L_4(2)$	$1/6,435$

The result now follows from direct calculation, using Proposition 8.0.57.

CHAPTER 9

Actions on 3-sets compared to actions on larger sets

In this chapter we obtain lower bounds on $\varepsilon_{3,k}(x)$ for $x \in S_n$ and $4 \le k \le n-4$ which we will need in the next two chapters. Namely, we will prove the following result.

LEMMA 9.0.64. *If $3 \le k \le n-3$ and $x \in S_n$ then*
$$\varepsilon_{3,k}(x) \ge -\frac{3}{(n-1)(n-3)}.$$

Lemma 9.0.64 follows immediately from Burnside's lemma and the following result.

LEMMA 9.0.65. *If $3 \le k \le n-3$ and $x \in S_n$ then*
$$\delta_{3,k}(x) \ge -\frac{3}{(n-1)(n-3)}.$$

PROOF. The proof uses induction on n in a similar manner to the proof of Lemma 6.0.52. To start the induction we note that the claim of the lemma certainly holds when $n \le 7$. Now assume that $n > 7$. Let $x \in S_n$ have shape $(\lambda_1, \ldots, \lambda_r)$. Arguing as we did in the proof of Lemma 6.0.52, we see that we may assume that $2 \le \lambda_1 \le 7$, and it follows that $r > 1$. Let $x^- \in S_{n-\lambda_1}$ have shape $(\lambda_2, \ldots, \lambda_r)$. If $3 \le k - \lambda_1 \le k \le n - \lambda_1 - 3$ then, using our inductive hypothesis, we have

$$\frac{f_k(x)}{\binom{n}{k}} = \frac{f_k(x^-)}{\binom{n-\lambda_1}{k}}\frac{\binom{n-\lambda_1}{k}}{\binom{n}{k}} + \frac{f_{k-\lambda_1}(x^-)}{\binom{n-\lambda_1}{k-\lambda_1}}\frac{\binom{n-\lambda_1}{k-\lambda_1}}{\binom{n}{k}}$$

$$\le \left[\frac{f_3(x^-)}{\binom{n-\lambda_1}{3}} + \frac{3}{(n-\lambda_1-1)(n-\lambda_1-3)}\right]\frac{\binom{n-\lambda_1}{k}+\binom{n-\lambda_1}{k-\lambda_1}}{\binom{n}{k}}$$

$$\le \left[\frac{f_3(x)}{\binom{n}{3}} + \frac{3\binom{n-\lambda_1}{3}}{(n-\lambda_1-1)(n-\lambda_1-3)\binom{n}{3}}\right]\frac{\binom{n}{3}}{\binom{n-\lambda_1}{3}}\frac{\binom{n-\lambda_1}{k}+\binom{n-\lambda_1}{k-\lambda_1}}{\binom{n}{k}}.$$

Now

(9.1) $$\frac{\binom{n}{3}}{\binom{n-\lambda_1}{3}}\frac{\binom{n-\lambda_1}{k}+\binom{n-\lambda_1}{k-\lambda_1}}{\binom{n}{k}} = \frac{\prod_{j=0}^{\lambda_1-1}(n-k-j) + \prod_{j=0}^{\lambda_1-1}(k-j)}{\prod_{j=0}^{\lambda_1-1}(n-3-j)}.$$

Since there is nothing to prove when $k \in \{3, n-3\}$, we may assume that $4 \le k \le n-4$. Under this assumption, the right side of inequality (9.1) is maximized when

$k = 4$, so

$$(9.2) \qquad \frac{\binom{n}{3}}{\binom{n-\lambda_1}{3}} \frac{\binom{n-\lambda_1}{k} + \binom{n-\lambda_1}{k-\lambda_1}}{\binom{n}{k}} \le \frac{\prod_{j=0}^{\lambda_1-1}(n-4-j) + \prod_{j=0}^{\lambda_1-1}(4-j)}{\prod_{j=0}^{\lambda_1-1}(n-3-j)}.$$

It is straightforward to check that the right side of inequality (9.2) is at most 1 for all pairs k, λ_1 which are currently under consideration (note that if $\lambda_1 = 2$ then $k \ge 5$ and so $n \ge 10$). So, we can prove the claim of the lemma under our current assumptions by showing that

$$(9.3) \qquad \frac{3}{(n-\lambda_1-1)(n-\lambda_1-3)} \frac{\binom{n-\lambda_1}{3}}{\binom{n}{3}} \le \frac{3}{(n-1)(n-3)}.$$

Now

$$\frac{3}{(n-\lambda_1-1)(n-\lambda_1-3)} \frac{\binom{n-\lambda_1}{3}}{\binom{n}{3}} = \frac{3(n-\lambda_1)(n-\lambda_1-2)}{n(n-1)(n-2)(n-\lambda_1-3)}.$$

Therefore, if we set

$$\phi(n, \lambda_1) := n(n-2)(n-\lambda_1-3) - (n-\lambda_1)(n-\lambda_1-2)(n-3)$$

then inequality (9.3) is equivalent to the inequality

$$(9.4) \qquad \phi(n, \lambda_1) \ge 0.$$

Under our current assumptions, we have $2 \le \lambda_1 \le n - 6$. If we fix n, then $\phi(n, \lambda)$ is quadratic in λ with negative leading coefficient, and is therefore bounded above by the max $\{\phi(n, 2), \phi(n, n-6)\}$. Now

$$\phi(n, 2) = 2(n-2)(n-6) > 0,$$

and

$$\phi(n, n-6) = 3(n-4)(n-6) > 0,$$

so inequality (9.4) holds as desired.

It remains to verify the claim of the lemma when k is *too big*, that is, $k > n - \lambda_1 - 3$ or *too small*, that is, $k - \lambda_1 < 3$, or both too big and too small. If k is both too big and too small then the triple (n, k, λ_1) is listed in the table below.

n	k	λ_1
18	9	7
17	8, 9	7
16	8	7, 6
16	7, 9	7
15	7, 8	7, 6
15	6, 9	7
14	7	7, 6, 5
14	6, 8	7, 6
14	5, 9	7
13	6, 7	7, 6, 5
13	5, 8	7, 6
13	4, 9	7
12	6	7, 6, 5, 4
12	5, 7	7, 6, 5
12	4, 8	7, 6
11	5, 6	7, 6, 5, 4
11	4, 7	7, 6, 5
10	5	7, 6, 5, 4, 3
10	4, 6	7, 6, 5, 4
9	4, 5	7, 6, 5, 4, 3
8	4	7, 6, 5, 4, 3, 2

For every (x, n, k), we have

(9.5)
$$\frac{f_k(x)}{\binom{n}{k}} \leq \frac{\binom{n-\lambda_1}{k} + \binom{n-\lambda_1}{k-\lambda_1}}{\binom{n}{k}}.$$

It is straightforward to show that in the cases described in the table above, the right side of inequality (9.5) is bounded above by $\frac{3}{(n-1)(n-3)}$ unless the shape of x is either $(\lambda_1, 1, \ldots, 1)$ or $(\lambda_1, 2, 1, \ldots, 1)$ and (n, k, λ_1) is one of the triples in the table below.

n	k	λ_1
14	6, 8	6
13	6, 7	5
13	4, 9	7
12	6	4
11	5, 6	4
11	4, 7	5
10	5	3
10	4, 6	4
9	4, 5	3, 4
8	4	3, 2

In each of these cases, direct computation shows that the claim of the lemma holds.

We are left with the case where k is either too big or too small but not both. As in the proof of Lemma 6.0.52, we may assume that k is too small but not too big, so $k - \lambda_1 < 3$ but $n - k - \lambda_1 \geq 3$. If $k < \lambda_1$ then

$$
\begin{aligned}
\frac{f_k(x)}{\binom{n}{k}} &= \frac{f_k(x^-)}{\binom{n}{k}} \\
&= \frac{f_k(x^-)}{\binom{n-\lambda_1}{k}} \frac{\binom{n-\lambda_1}{k}}{\binom{n}{k}} \\
&\leq \left[\frac{f_3(x^-)}{\binom{n-\lambda_1}{3}} + \frac{3}{(n-\lambda_1-1)(n-\lambda_1-3)}\right] \frac{\binom{n-\lambda_1}{k}}{\binom{n}{k}} \\
&\leq \left[\frac{f_3(x)}{\binom{n}{3}} + \frac{3\binom{n-\lambda_1}{3}}{(n-\lambda_1-1)(n-\lambda_1-3)\binom{n}{3}}\right] \frac{\binom{n}{3}}{\binom{n-\lambda_1}{3}} \frac{\binom{n-\lambda_1}{k}}{\binom{n}{k}},
\end{aligned}
$$

the first inequality following from the induction hypothesis. As noted above in inequality (9.3), we have

$$
\frac{3}{(n-\lambda_1-1)(n-\lambda_1-3)} \frac{\binom{n-\lambda_1}{3}}{\binom{n}{3}} \leq \frac{3}{(n-1)(n-3)}.
$$

Also,

$$
\frac{\binom{n}{3}}{\binom{n-\lambda_1}{3}} \frac{\binom{n-\lambda_1}{k}}{\binom{n}{k}} = \prod_{j=0}^{\lambda_1-1} \frac{n-k-j}{n-3-j} \leq 1,
$$

so the claim of the lemma holds in this case.

If $\lambda_1 = k$ then we have

$$
f_k(x) = 1 + f_k(x^-),
$$

and arguing as we did above we get

$$
\frac{f_k(x)}{\binom{n}{k}} \leq \frac{1}{\binom{n}{k}} + \left[\frac{f_3(x)}{\binom{n}{3}} + \frac{3}{(n-1)(n-3)}\right] \frac{\binom{n}{3}}{\binom{n-k}{3}} \frac{\binom{n-k}{k}}{\binom{n}{k}}.
$$

Therefore, the claim of the lemma holds if the inequality

$$
(9.6) \qquad \frac{1}{\binom{n}{k}} \leq \left[\frac{f_3(x)}{\binom{n}{3}} + \frac{3}{(n-1)(n-3)}\right] \left[1 - \frac{\binom{n}{3}\binom{n-k}{k}}{\binom{n-k}{3}\binom{n}{k}}\right]
$$

holds. Since $f_3(x) \geq 0$ for all x, inequality (9.6) holds if the inequality

$$
(9.7) \qquad 1 \leq \frac{3}{(n-1)(n-3)} \left[\binom{n}{k} - \frac{\binom{n}{3}\binom{n-k}{k}}{\binom{n-k}{3}}\right]
$$

holds. Since there is nothing to prove if $k = 3$, we may assume that $4 \leq k \leq 7$. Given our assumption that $n - \lambda_1 - k \geq 3$, it is straightforward to show that inequality (9.7) holds for each possible k.

Say $k = \lambda_1 + 1$. Then

$$
f_k(x) = f_1(x^-) + f_k(x^-) = f_1(x) + f_k(x^-),
$$

so
$$\frac{f_k(x)}{\binom{n}{k}} = \frac{f_1(x)}{\binom{n}{k}} + \frac{f_k(x^-)}{\binom{n-k+1}{k}} \frac{\binom{n-k+1}{k}}{\binom{n}{k}}$$
$$\leq \frac{f_1(x)}{\binom{n}{k}} + \left[\frac{f_3(x^-)}{\binom{n-k+1}{3}} + \frac{3}{(n-k)(n-k-2)}\right]\frac{\binom{n-k+1}{k}}{\binom{n}{k}}$$
$$\leq \frac{f_1(x)}{\binom{n}{k}} + \frac{f_3(x)}{\binom{n}{3}} \frac{\binom{n}{3}}{\binom{n-k+1}{3}} \frac{\binom{n-k+1}{k}}{\binom{n}{k}} + \frac{3}{(n-k)(n-k-2)}\frac{\binom{n-k+1}{k}}{\binom{n}{k}}.$$

Fix n and define
$$\phi_1(k) := \frac{1}{\binom{n}{k}},$$
$$\phi_2(k) := \frac{\binom{n}{3}}{\binom{n-k+1}{3}} \frac{\binom{n-k+1}{k}}{\binom{n}{k}}$$

and
$$\phi_3(k) := \frac{3}{(n-k)(n-k-2)}\frac{\binom{n-k+1}{k}}{\binom{n}{k}}.$$

Certainly $\phi_1(k)$ decreases as k increases from 4 to $\lfloor \frac{n}{2} \rfloor$. We have
$$\frac{\phi_2(k+1)}{\phi_2(k)} = \frac{(n-2k+1)(n-2k)}{(n-k)(n-k-2)},$$

and
$$\frac{\phi_3(k+1)}{\phi_3(k)} = \frac{(n-k-2)(n-2k+1)(n-2k)}{(n-k+1)(n-k-1)(n-k-3)},$$

so $\phi_2(k)$ and $\phi_3(k)$ also decrease as k increases from 4 to $\lfloor \frac{n}{2} \rfloor$. This gives
$$\frac{f_k(x)}{\binom{n}{k}} \leq \frac{f_1(x)}{\binom{n}{4}} + \frac{f_3(x)}{\binom{n}{3}} \frac{\binom{n}{3}}{\binom{n-3}{3}} \frac{\binom{n-3}{4}}{\binom{n}{4}} + \frac{3}{(n-4)(n-6)}\frac{\binom{n-3}{4}}{\binom{n}{4}}.$$

It follows that the claim of the lemma holds if the inequality
(9.8) $$8f_1(x) - (6n-15) \leq 6f_3(x)$$
holds. Since we are assuming that $n > 7$, we have $6n - 15 \geq 33$, so inequality (9.8) certainly holds if $f_1(x) \leq 4$. If $f_1(x) \geq 5$ then
$$\frac{f_3(x)}{f_1(x)} \geq \frac{\binom{f_1(x)}{3}}{f_1(x)} = \frac{(f_1(x)-1)(f_1(x)-2)}{6} \geq 2,$$

and again inequality (9.8) holds, so the lemma holds in this case.

We are left with the case $k = \lambda_1 + 2$. We handle the case $k = 4$ separately from the others. So, if $k = 4$ then x has shape $1^a 2^{\frac{n-a}{2}}$. Since k is not too big, we have $n \geq 9$. Now
$$f_3(x) = \binom{a}{3} + \frac{a(n-a)}{2},$$

and
$$f_4(x) = \binom{a}{4} + \binom{a}{2}\frac{n-a}{2} + \binom{(n-a)/2}{2}.$$

Direct calculation gives
$$\delta_{3,4}(x) = \frac{(n-a)(a-1)(3n+a(a-8)-6)}{n(n-1)(n-2)(n-3)}.$$

It is straightforward to show that under our assumption $n \geq 9$ that if $a \geq 2$ then
$$3n + a(a-8) - 6 > 0$$
so $\delta_{3,4}(x) > 0$. Also, $\delta_{3,4}(x) = -\frac{3}{(n-1)(n-3)}$ when $a = 0$ and $\delta_{3,4}(x) = 0$ when $a = 1$, so the claim of the lemma holds in this case.

We may now assume that $k \geq 5$, so $\lambda_1(x) \geq 3$. We have
$$f_k(x) = f_k(x^-) + f_2(x^-) = f_k(x^-) + f_2(x).$$
Arguing as we did in the case $k = \lambda_1 + 1$, we get
$$\frac{f_k(x)}{\binom{n}{k}} = \frac{f_2(x)}{\binom{n}{k}} + \frac{f_k(x^-)}{\binom{n}{k}}$$
$$\leq \frac{f_2(x)}{\binom{n}{k}} + \left[\frac{f_3(x)}{\binom{n}{3}}\frac{\binom{n}{3}}{\binom{n-k+2}{3}} + \frac{3}{(n-k+1)(n-k-1)}\right]\frac{\binom{n-k+2}{k}}{\binom{n}{k}}.$$
Again fix n and define
$$\psi_1(k) := \frac{1}{\binom{n}{k}},$$
$$\psi_2(k) := \frac{\binom{n}{3}}{\binom{n-k+2}{3}}\frac{\binom{n-k+2}{k}}{\binom{n}{k}},$$
and
$$\psi_3(k) := \frac{3}{(n-k+1)(n-k-1)}\frac{\binom{n-k+2}{k}}{\binom{n}{k}}.$$
Then
$$\frac{\psi_2(k+1)}{\psi_2(k)} = \frac{(n-2k+2)(n-2k+1)}{(n-k)(n-k-1)} < 1$$
and
$$\frac{\psi_3(k+1)}{\psi_3(k)} = \frac{(n-k+1)(n-k-1)(n-2k+2)(n-2k+1)}{(n-k+2)(n-k)^2(n-k-2)} < 1.$$
Therefore, each of $\psi_1(k)$, $\psi_2(k)$ and $\psi_3(k)$ decreases as k increases from 5 to $\lfloor \frac{n}{2} \rfloor$. This gives
$$\frac{f_k(x)}{\binom{n}{k}} \leq \frac{f_2(x)}{\binom{n}{5}} + \left[\frac{f_3(x)}{\binom{n-3}{3}} + \frac{3}{(n-4)(n-6)}\right]\frac{\binom{n-3}{5}}{\binom{n}{5}}.$$
It follows that the claim of the lemma holds if the inequality
(9.9) $\qquad 40f_2(x) \leq (12n - 60)f_3(x) + 9n^2 - 63n + 105$
holds. Let $a = f_1(x)$. Then
$$f_2(x) \leq \binom{a}{2} + \frac{n-a}{2},$$
and
$$f_3(x) \geq \binom{a}{3}.$$
It follows that inequality (9.9) holds if the inequality
(9.10) $\qquad 2a(a-2)\left[(a-1)(n-5) - 10\right] + 9n^2 - 83n + 105 \geq 0$
holds. Since we are assuming that $k \geq 5$ and k is not too big, we have $n \geq 11$. It follows that inequality (9.10) holds for all a, and this completes the proof of the lemma. $\square$

Note that in the proof of Lemma 9.0.65 we saw that if x has shape $2^{\frac{n}{2}}$ then $\delta_{3,4}(x) = -\frac{3}{(n-1)(n-3)}$, so the bound given in the lemma is the best one possible.

CHAPTER 10

A transposition and an n-cycle

In this chapter we examine the case where E contains three elements, one of which is a transposition and one of which is an n-cycle. The results we obtain will be used in Chapters 11, 12 and 13.

Say $E = (t, z, y)$ where t has shape $1^{n-2}2^1$, z has shape (n) and $y = (tz)^{-1}$. We assume without loss of generality that

- $z = (1, 2, \ldots, n)$, and
- $t = (1, a+1)$ for some $a \in [n-1]$.

Now

- $y^{-1} = (1, a+2, a+3, \ldots, n)(2, 3, \ldots, a+1)$,

so y has shape $(a, n-a)$.

Certainly $\langle t, y, z \rangle$ is transitive. If

$$\Pi = [\pi_1 | \ldots | \pi_d]$$

is a z-invariant partition of $[n]$ then $d | n$ and we may assume that

$$\pi_i = \{j \in [n] : j \equiv i \bmod d\}.$$

It follows that Π is t-invariant if and only if $d | a$. Therefore, $\langle t, y, z \rangle$ is primitive if and only if $\gcd(a, n) = 1$. If this last condition holds then, since t is a transposition, we have $\langle t, y, z \rangle = S_n$ (see for example [**DiMo**, Theorem 3.3A]).

So, we assume from now on that $\gcd(a, n) = 1$ and we set $b = n - a$. We may assume that $a > b$, so y has shape (a, b). We will obtain an explicit formula for $g_k(E)$ from which the next theorem will follow.

THEOREM 10.0.66. *Assume that $t \in S_n$ has shape $1^{n-2}2^1$, y has shape (a, b) with $\gcd(a, n) = 1$, z has shape (n) and $tzy = 1$. Let $E = (t, z, y)$. Then*

(1) *We have $g_1(E) = g_2(E) = 0$.*
(2) *There exists some constant $c > 0$ such that if $n > 7$ and $3 \leq k \leq n - 3$ then*

$$\frac{g_k(E)}{\binom{n}{k}} > \frac{c}{n}$$

(3) If $n > 6$ then $g_3(E) > 2$ unless one of the conditions in the table below holds.

n	a	$g_3(E)$
7	6	0
7	5	1
7	4	1
8	7	1
8	5	2
9	8	1
9	7	2
10	9	2

(4) If $n > 9$ then $g_4(E) - g_3(E) > 2$.

PROOF. Note first that

- $i_k(t) = \binom{n-2}{k-1}$

by Lemma 2.0.19. For each $j \in [n]$, z^j is homocyclic of order $\frac{n}{\gcd(n,j)}$. It now follows from Burnside's lemma that if we let $\phi(m)$ be Euler's function (that is, the number of units in $\mathbb{Z}_m$) and define

$$P_k(n) := \sum_{1 < d \mid \gcd(k,n)} \phi(d) \binom{n/d}{k/d}$$

then

$$o_k(z) = \frac{1}{n}\left[\binom{n}{k} + P_k(n)\right],$$

so

- $i_k(z) = \frac{n-1}{n}\binom{n}{k} - \frac{1}{n}P_k(n)$.

Write $y = y_a y_b$, where y_a is an a-cycle and y_b is a b-cycle. Set $S_a = supp(y_a)$, $S_b = supp(y_b)$. For $X \subseteq [n]$ let $X_a = X \cap S_a$, $X_b = X \cap S_b$. Let $X\langle y \rangle$ be the $\langle y \rangle$-orbit containing X and define $X_a \langle y_a \rangle$ and $X_b \langle y_b \rangle$ similarly. Set

$$\mathcal{X} := \{A \cup B : A \in X_a \langle y_a \rangle, B \in X_b \langle y_b \rangle\}.$$

Certainly $X\langle y \rangle \subseteq \mathcal{X}$. Since $\gcd(a,b) = 1$ we have $\langle y \rangle = \langle y_a \rangle \times \langle y_b \rangle$, and every subgroup of $\langle y \rangle$ is of the form $G \times H$ with $G \leq \langle y_a \rangle$ and $H \leq \langle y_b \rangle$. Therefore

$$Stab_{\langle y \rangle}(X) = Stab_{\langle y_a \rangle}(X_a) \times Stab_{\langle y_b \rangle}(X_b).$$

Simple computation now gives $|X\langle y \rangle| = |\mathcal{X}|$, so

$$X\langle y \rangle = \mathcal{X}.$$

Therefore, if we define $o'_l(y_a)$ to be the number of orbits of $\langle y_a \rangle$ on l-subsets of $[a]$ and define $o'_l(y_b)$ similarly then we have

$$o_k(y) = \sum_{j=0}^{k} o'_j(y_a) o'_{k-j}(y_b).$$

Applying our result for $o_k(z)$, we get

$$o_k(y) = \frac{1}{a}\left[\binom{a}{k} + P_k(a)\right] + \frac{1}{b}\left[\binom{b}{k} + P_k(b)\right]$$
$$+ \frac{1}{ab}\sum_{j=1}^{k-1}\left[\binom{a}{j} + P_j(a)\right]\left[\binom{b}{k-j} + P_{k-j}(b)\right].$$

We have $i_1(t) = 1$, $i_1(z) = n-1$ and $i_1(y) = n-2$, so $g_1(E) = 0$. We have $i_2(t) = n-2$, and using

$$P_2(n) = \begin{cases} \frac{n}{2} & \text{if } n \text{ is even,} \\ 0 & \text{if } n \text{ is odd,} \end{cases}$$

we get

$$i_2(y) = \begin{cases} \binom{n}{2} - \frac{n}{2} & n \text{ even,} \\ \binom{n}{2} - \frac{n+1}{2} & n \text{ odd,} \end{cases}$$

and

$$i_2(z) = \begin{cases} \binom{n}{2} - \frac{n}{2} & n \text{ even,} \\ \binom{n}{2} - \frac{n-1}{2} & n \text{ odd.} \end{cases}$$

It follows that $g_2(E) = 0$.

To determine $g_3(E)$ and $g_4(E)$ we must calculate the polynomials $P_k(n)$ for $2 \le k \le 4$. $P_2(n)$ was described above, and we have

$$P_3(n) = \begin{cases} \frac{2n}{3} & 3|n, \\ 0 & \text{otherwise,} \end{cases}$$

and

$$P_4(n) = \begin{cases} \binom{n/2}{2} + 2\frac{n}{4} & \text{if } n \equiv 0 \bmod 4, \\ \binom{n/2}{2} & \text{if } n \equiv 2 \bmod 4, \\ 0 & \text{otherwise.} \end{cases}$$

This gives

$$i_3(t) = \frac{n^2 - 5n + 6}{2},$$

$$i_3(y) = \begin{cases} \binom{n}{3} - \frac{n^2+2-2a(n-a)}{6} & 2|n, 3|a(n-a), \\ \binom{n}{3} - \frac{n^2-2-2a(n-a)}{6} & 2|n, 3\nmid a(n-a), \\ \binom{n}{3} - \frac{n^2+5-2a(n-a)}{6} & 2\nmid n, 3|a(n-a), \\ \binom{n}{3} - \frac{n^2+1-2a(n-a)}{6} & 2\nmid n, 3\nmid a(n-a), \end{cases}$$

and

$$i_3(z) = \begin{cases} \binom{n}{3} - \frac{n^2-3n+6}{6} & 3|n, \\ \binom{n}{3} - \frac{n^2-3n+2}{6} & 3\nmid n. \end{cases}$$

Straightforward calculations now give the results in the table below.

$(n \bmod 6, a \bmod 6)$	$g_3(E)$
$(1,0), (1,1), (1,3), (1,4),$ $(3,1), (3,2), (3,4), (3,5),$ $(5,0), (5,2), (5,3), (5,5)$	$\frac{1}{12}\left[n^2 - 12n + 23 + 2a(n-a)\right]$
$(0,1), (0,5), (2,3), (2,5), (4,1), (4,3)$	$\frac{1}{12}\left[n^2 - 12n + 26 + 2a(n-a)\right]$
$(1,2), (1,5), (5,1), (5,4)$	$\frac{1}{12}\left[n^2 - 12n + 27 + 2a(n-a)\right]$
$(2,1), (4,5)$	$\frac{1}{12}\left[n^2 - 12n + 30 + 2a(n-a)\right]$

It is now straightforward to prove part 3 of the theorem, using the fact that $a(n-a)$ is minimized by taking a to be as large as possible given the constraints on (n,a) in each case. Also, we have

$$\lim_{n\to\infty} \min_a \frac{ng_3(E)}{\binom{n}{3}} = \frac{1}{2},$$

and since $g_3(E) > 0$ for $n > 7$ there is some constant c' such that

$$\frac{g_3(E)}{\binom{n}{3}} > \frac{c'}{n}$$

whenever $n > 7$. By Lemma 2.0.12 we have $g_k(E) > 0$ whenever $4 \leq k \leq n-4$, and by Proposition 2.0.6 and Lemma 9.0.64 we have

$$\frac{g_k(E)}{\binom{n}{k}} - \frac{g_3(E)}{\binom{n}{3}} \geq \frac{1}{\binom{n}{k}} - \frac{1}{\binom{n}{3}} - \frac{9}{2(n-1)(n-3)}$$

for $4 \leq k \leq n-4$. Part 2 of the theorem follows easily.

We have

$$i_4(t) = \frac{n^3 - 9n^2 + 26n - 24}{6}$$

and

$$i_4(z) = \begin{cases} \binom{n}{4} - \frac{n^3 - 6n^2 + 14n}{24} & n \equiv 0 \bmod 4, \\ \binom{n}{4} - \frac{n^3 - 6n^2 + 14n - 12}{24} & n \equiv 2 \bmod 4, \\ \binom{n}{4} - \frac{n^3 - 6n^2 + 11n - 6}{24} & n \equiv 1,3 \bmod 4. \end{cases}$$

If we set

$$\begin{aligned} Q &= 0, \\ R &= 16, \\ S &= 3n + 3a + 4, \\ T &= 3n + 3a, \\ U &= 3n + 3a + 16, \\ V &= 3n + 3a - 12, \\ W &= 6n - 3a + 16, \\ X &= 6n - 3a - 12, \\ Y &= 6n - 3a + 4, \\ Z &= 6n - 3a \end{aligned}$$

then

$$i_4(y) = \binom{n}{4} - \frac{1}{24}\left[n^3 - 2n^2 - 7n + 10 - a(n-a)(3n-10) + D\right],$$

where D takes one of the values $Q, \ldots, Z$ as determined by the pair $(n \bmod 12, a \bmod 12)$. The value taken by D is given in the table below. The rows are determined by $n \bmod 12$ and the columns are determined by $a \bmod 12$. The symbol $-$ in an entry means that we cannot have $\gcd(n,a) = 1$ for the given values.

	0	1	2	3	4	5	6	7	8	9	10	11
0	–	Q	–	–	–	Q	–	Q	–	–	–	Q
1	W	U	X	S	W	T	Y	S	Z	U	Y	V
2	–	Q	–	R	–	R	–	Q	–	R	–	R
3	–	V	X	–	Z	V	–	T	Z	–	X	T
4	–	R	–	R	–	Q	–	R	–	R	–	Q
5	W	T	Y	S	Z	U	Y	V	W	U	X	S
6	–	Q	–	–	–	Q	–	Q	–	–	–	Q
7	W	S	X	U	W	V	Y	U	Z	S	Y	T
8	–	Q	–	R	–	R	–	Q	–	R	–	R
9	–	T	X	–	Z	T	–	V	Z	–	X	V
10	–	R	–	R	–	Q	–	R	–	R	–	Q
11	W	V	Y	U	Z	S	Y	T	W	S	X	U

It now follows that if we define

$$\begin{aligned} Q' &= 3n - 6, \\ Q'' &= 3n + 6, \\ R' &= 3n + 10, \\ R'' &= 3n + 22 \end{aligned}$$

then

$$g_4(E) = \frac{1}{48}\left[2n^3 - 28n^2 + 100n - 52 + a(n-a)(3n-10) - D'\right],$$

where D' takes one of the values $Q', Q'', R', R'', S, \ldots, Z$. The table below, which is formatted in the same way as the table above, indicates which value D' takes.

	0	1	2	3	4	5	6	7	8	9	10	11
0	–	Q''	–	–	–	Q''	–	Q''	–	–	–	Q''
1	W	U	X	S	W	T	Y	S	Z	U	Y	V
2	–	Q'	–	R'	–	R'	–	Q'	–	R'	–	R'
3	–	V	X	–	Z	V	–	T	Z	–	X	T
4	–	R''	–	R''	–	Q''	–	R''	–	R''	–	Q''
5	W	T	Y	S	Z	U	Y	V	W	U	X	S
6	–	Q'	–	–	–	Q'	–	Q'	–	–	–	Q'
7	W	S	X	U	W	V	Y	U	Z	S	Y	T
8	–	Q''	–	R''	–	R''	–	Q''	–	R''	–	R''
9	–	T	X	–	Z	T	–	V	Z	–	X	V
10	–	R'	–	R'	–	Q'	–	R'	–	R'	–	Q'
11	W	V	Y	U	Z	S	Y	T	W	S	X	U

This gives

$$g_4(E) - g_3(E) = \frac{1}{48}\left[2n^3 - 32n^2 + 148n - 172 + a(n-a)(3n-18) - D' + C'\right],$$

where $C' \in \{0, 12, 16, 28\}$ is determined by the pair $(n \bmod 6, a \bmod 6)$. Part 4 of the theorem follows from straightforward calculations. $\square$

CHAPTER 11

Asymptotic behavior of $g_k(E)$

In this chapter we prove Theorem 1.3.1. First we describe for each k a class $\mathcal{E}(k) = \{E_n(k) : n > k\}$ such that $|E_n(k)| = 5$ for all n and there is some constant $c_k > 0$ with

(11.1) $$\lim_{n \to \infty} \frac{g_k(E_n(k))}{\frac{1}{n}\binom{n}{k}} = c_k,$$

thereby proving part (2) of the theorem. As we will see, we have $c_k \geq \frac{3}{2}$ for all k. These examples all are derived from a particular case from the class of examples examined in Chapter 10. Namely, for a positive integer n, let $z = (1, 2, \ldots, n)$ be an n-cycle and let $t = (1, 2)$. Let $y = (tz)^{-1}$, so y has shape $(n-1, 1)$. Let $D_n = (t, z, y)$. As is well known, we have $\langle D_n \rangle = S_n$. We saw in Chapter 10 that

$$g_2(D_n) = 0$$

for all n. In other words, by the Riemann-Hurwitz formula, we have

(11.2) $$i_2(t) + i_2(y) + i_2(z) = 2\left[\binom{n}{2} - 1\right].$$

Lemma 2.0.19 says that

(11.3) $$i_2(t) = n - 2.$$

For $n > 2$ set

$$E_n(2) := (t, t, t, z, y).$$

Certainly $tttzy = 1$ and $\langle E_n(2) \rangle = S_n$. The Riemann-Hurwitz formula, along with equations (11.2) and (11.3) give

$$2\left[g_2(E_n(2)) + \binom{n}{2} - 1\right] = 2\left[\binom{n}{2} - 1\right] + 2(n-2).$$

Therefore

$$g_2(E_n(2)) = n - 2$$

for all n, and equation (11.1) holds with $c_2 = 2$.

Now assume $k > 2$. Let $u, v, w, x \in S_n$ be involutions with $uv = z$ and $wx = y$. Set

$$E_n(k) := (t, u, v, w, x).$$

As noted in Chapter 5, it is clear that $tuvwx = 1$ and $\langle E_n(k) \rangle = S_n$. Also, if n is even then we may assume that u has shape $2^{\frac{n}{2}}$ and v, w, x all have shape $1^2 2^{\frac{n-2}{2}}$

while if n is odd then we may assume that x has shape $1^3 2^{\frac{n-3}{2}}$ and u,v,w all have shape $1^1 2^{\frac{n-1}{2}}$. the possible indices for u,v,w,x are listed in the table below.

$shape(y)$	$i_k(y)$ (k odd)	$i_k(y)$ (k even)
$2^{\frac{n}{2}}$	$\frac{1}{2}\binom{n}{k}$	$\frac{1}{2}\left[\binom{n}{k} - \binom{n/2}{k/2}\right]$
$1^1 2^{\frac{n-1}{2}}$	$\frac{1}{2}\left[\binom{n}{k} - \binom{(n-1)/2}{(k-1)/2}\right]$	$\frac{1}{2}\left[\binom{n}{k} - \binom{(n-1)/2}{k/2}\right]$
$1^2 2^{\frac{n-2}{2}}$	$\frac{1}{2}\left[\binom{n}{k} - 2\binom{(n-1)/2}{(k-1)/2}\right]$	$\frac{1}{2}\left[\binom{n}{k} - \binom{n/2}{k/2}\right]$
$1^3 2^{\frac{n-3}{2}}$	$\frac{1}{2}\left[\binom{n}{k} - 3\binom{(n-3)/2}{(k-1)/2} - \binom{(n-3)/2}{(k-3)/2}\right]$	$\frac{1}{2}\left[\binom{n}{k} - \binom{(n-3)/2}{k/2} - 3\binom{(n-3)/2}{(k-2)/2}\right]$

We can now calculate $g_k(E_n(k))$ directly for all appropriate pairs n, k to get the results in the following table.

$n \bmod 2$	$k \bmod 2$	$g_k(E_n(k))$
0	0	$\frac{1}{2}\binom{n-2}{k-1} - \binom{n/2}{k/2} + 1$
0	1	$\frac{1}{2}\binom{n-2}{k-1} - \frac{3}{2}\binom{(n-2)/2}{(k-1)/2} + 1$
1	0	$\frac{1}{2}\binom{n-2}{k-1} - \binom{(n-1)/2}{k/2} - \frac{1}{2}\binom{(n-3)/2}{(k-2)/2} + 1$
1	1	$\frac{1}{2}\binom{n-2}{k-1} - \binom{(n-1)/2}{(k-1)/2} - \frac{1}{2}\binom{(n-3)/2}{(k-1)/2} + 1$

It follows that for each $k > 2$ we have

$$\lim_{n\to\infty} \frac{g_k(E_n(k))}{\frac{1}{n}\binom{n}{k}} = \frac{k}{2}.$$

Now we prove part (1) of Theorem 1.3.1. Since we are assuming that n is large, we may assume that one of conditions (1), (2) of Theorem 4.0.30 holds. Assume first that condition (1) holds.

Fix E as described in the theorem. We know that $g_2(E) > cn$ with c as in Theorem 4.0.30, so

$$\frac{g_2(E)}{\binom{n}{2}} > \frac{2c}{n}.$$

Now Lemma 2.0.6 gives

$$\frac{g_k(E)}{\binom{n}{k}} > \frac{2c}{n} - \frac{2}{n(n-1)} + \frac{1}{2}\sum_{x\in E}\varepsilon_{2,k}(x).$$

Set

$$\mathcal{F} := \{x \in E : \varepsilon_{2,k}(x) < 0\}.$$

By Lemma 6.0.54, we have

$$\frac{g_k(E)}{\binom{n}{k}} > \frac{2c}{n} - \frac{2}{n(n-1)} - \frac{|\mathcal{F}|}{(n-1)(n-2)}.$$

In particular, if $|\mathcal{F}| \leq 5$ then

$$\frac{g_k(E)}{\binom{n}{k}} > \frac{2c}{n} - \frac{2}{n(n-1)} - \frac{5}{(n-1)(n-2)}.$$

Since $g_k(E) > 0$ for all E under consideration, we can find some C such that

(11.4) $$\frac{g_k(E)}{\binom{n}{k}} > \frac{C}{n},$$

whenever $|\mathcal{F}| \leq 5$.

On the other hand, assuming as we may that $n > 8$, if $x \in \mathcal{F}$ then $f_1(x) \leq 1$. By Lemma 5.0.41, we have
$$\frac{i_2(x)}{\binom{n}{2}} \geq \frac{1}{2} - \frac{1}{2(n-1)}.$$
Now Lemma 6.0.54 gives
$$\frac{i_k(x)}{\binom{n}{k}} \geq \frac{1}{2} - \frac{1}{2(n-1)} - \frac{2}{(n-1)(n-2)}$$
$$= \frac{n(n-4)}{2(n-1)(n-2)},$$
Thus if $n > 20$, we have
$$2(g_k(E) - 1) \geq \left(\frac{5}{12}|\mathcal{F}| - 2\right)\binom{n}{k},$$
so if $|\mathcal{F}| \geq 6$ we have
$$\frac{g_k(E)}{\binom{n}{k}} > \frac{1}{4},$$
and again there is some C which can be chosen so that inequality (11.4) is satisfied.

On the other hand, if condition (2) of Theorem 4.0.30 holds but $k > 2$ then we have shown already that
$$\lim_{n \to \infty} \frac{g_k(E_n(k))}{\frac{1}{n}\binom{n}{k}} = \frac{k}{2}.$$

CHAPTER 12

An n-cycle - the proof of Theorem 1.2.1

In this chapter, we prove Theorem 1.2.1. The proof is similar to that of Theorem 1.1.2. We begin by finding lower bounds on $g_2(E) - g_1(E)$ and $g_3(E) - g_2(E)$ and we then use these bounds and Lemma 2.0.13 to show that any counterexample to the theorem involves the action of S_n or A_n on the cosets of a 3-homogeneous primitive subgroup. Then Theorem 4.0.33 is invoked to finish the proof. We begin with an examination of $g_2(E) - g_1(E)$.

THEOREM 12.0.67. *Let $E = (x_1, \ldots, x_r)$ be an r-tuple of nonidentity elements of S_n. Assume that $n > 20$, that $A_n \le \langle x_1, \ldots, x_r \rangle$ and that $\prod_{i=1}^r x_i = 1$. Assume also that x_r has cycle shape (n). Then there is some constant $c > 0$ such that exactly one of the following conditions holds.*
  (1) *We have $g_2(E) - g_1(E) \ge \max\{cn, 1\}$. Moreover, $g_2(E) - \frac{r}{3} > 0$, and if $r > 3$ or $n > 48$ then $g_2(E) - g_1(E) > 2$.*
  (2) *We have $g_1(E) = g_2(E) = 0$, and one of the following conditions holds.*
    (a) *$E \setminus \{x_r\}$ consists of one element of shape $1^{n-2}2^1$, one element of shape $1^3 2^{\frac{n-3}{2}}$ and one element of shape $1^1 2^{\frac{n-1}{2}}$.*
    (b) *$E \setminus \{x_r\}$ consists of one element of shape $1^{n-2}2^1$ and two elements of shape $1^2 2^{\frac{n-2}{2}}$.*
    (c) *$E \setminus \{x_r\}$ consists of one element of shape $1^{n-2}2^1$ and one element of shape $a^1 b^1$ with $\gcd(a, b) = 1$.*
    (d) *$E \setminus \{x_r\}$ consists of one element of shape $1^3 2^{\frac{n-3}{2}}$ and one element of shape $1^1 2^{\frac{n-5}{2}} 4^1$.*
    (e) *$E \setminus \{x_r\}$ consists of one element of shape $1^3 2^{\frac{n-3}{2}}$ and one element of shape $2^{\frac{n-3}{2}} 3^1$.*
    (f) *$E \setminus \{x_r\}$ consists of one element of shape $1^2 2^{\frac{n-2}{2}}$ and one element of shape $1^2 2^{\frac{n-6}{2}} 4^1$.*
    (g) *$E \setminus \{x_r\}$ consists of one element of shape $1^2 2^{\frac{n-2}{2}}$ and one element of shape $1^1 2^{\frac{n-4}{2}} 3^1$.*
    (h) *$E \setminus \{x_r\}$ consists of one element of shape $1^1 2^{\frac{n-1}{2}}$ and one element of shape $1^3 2^{\frac{n-7}{2}} 4^1$.*
    (i) *$E \setminus \{x_r\}$ consists of one element of shape $1^1 2^{\frac{n-1}{2}}$ and one element of shape $1^2 2^{\frac{n-5}{2}} 3^1$.*

Moreover, for each odd n and each condition 2(a),(c),(d),(e),(h),(i) and each even n and each condition 2(b),(c),(f),(g) there exists some $E \subset S_n$ satisfying the given condition.

PROOF. Note first that if $r \ge 5$ then the claim of the theorem follows immediately from Theorem 1.1.2.

Say $r = 4$. If n is even there exist $s, t \in S_n$ with shapes $1^2 2^{\frac{n-2}{2}}$ and $2^{\frac{n}{2}}$, respectively, such that $st = x_4$. If n is odd, there exist $s, t \in S_n$ both with shape $1^1 2^{\frac{n-1}{2}}$ such that $st = x_4$. Set
$$E' := (x_1, x_2, x_3, s, t).$$
Note that $\langle E' \rangle \geq \langle E \rangle$ and that $x_1 x_2 x_3 s t = 1$. If n is even then Burnside's Lemma gives
$$o_2(s) = o_2(t) = \frac{1}{2}\left[\binom{n}{2} + \frac{n}{2}\right],$$
while if n is odd then
$$o_2(s) = o_2(t) = \frac{1}{2}\left[\binom{n}{2} + \frac{n-1}{2}\right].$$
In either case, we have
$$o_2(s) + o_2(t) = \binom{n}{2} + \lfloor\frac{n}{2}\rfloor = \binom{n}{2} + o_2(z).$$
Therefore,
$$i_2(s) + i_2(t) = i_2(z),$$
and it follows that
$$g_2(E') = g_2(E).$$
Since $|E'| = 5$, it follows from Theorem 4.0.30 that one of 1, 2(a) or 2(b) holds. Moreover, we can obtain systems which satisfies 2(a) or 2(b) by taking systems $E' \subset S_n$ which satisfy condition 2(a) or 2(b) of Theorem 4.0.30, respectively, and appropriately combining two involutions from E' to obtain an n-cycle, as shown in Proposition 3.0.24.

Say $r = 3$. We will examine an exhaustive list of cases and exhibit in each case for which condition 2 does not hold a lower bound on $g_2(E) - g_1(E)$ which is linear or quadratic in n. The other lower bounds on $g_2(E) - g_1(E)$ and $g_2(E) - \frac{r}{3}$ claimed in the theorem follow immediately. Let $E = (x, y, z)$, where z has shape (n). Lemma 5.0.40(8a) gives
$$\varepsilon_2(z) = \begin{cases} -\frac{1}{n(n-1)} & n \text{ even}, \\ 0 & n \text{ odd}. \end{cases}$$
Note that one of x, y is not an involution. If $g_1(E) > 1$ then equation (5.1) of Lemma 5.0.43 and Lemma 5.0.40(8a,9) give
$$\begin{aligned} g_2(E) - g_1(E) &\geq \frac{n-3}{2} - \frac{n(n-1)}{4}\left(\frac{1}{2(n-1)} + \frac{1}{4(n-1)} + \frac{1}{n(n-1)}\right) \\ &= \frac{5n - 28}{16}. \end{aligned}$$
Say $g_1(E) = 1$. Equation (5.1) gives
$$g_2(E) - g_1(E) = \frac{n(n-1)}{4}\left(\varepsilon_2(x) + \varepsilon_2(y) + \varepsilon_2(z)\right).$$
Also,
$$o_1(x) + o_1(y) = n - 1.$$
We may (and do) assume that
$$o_1(x) \geq \frac{n-1}{2}.$$

12. AN n-CYCLE

Then $o_1(y) \leq \frac{n-1}{2}$. By Lemma 5.0.40(8a,9) we have
$$\varepsilon_2(y) \geq -\frac{1}{4(n-1)}.$$
If $f_1(x) \geq 2$ then Lemma 5.0.40(1-6) gives
$$\varepsilon_2(x) \geq \frac{n-2}{2n(n-1)},$$
so
$$g_2(E) - g_1(E) \geq \frac{n-8}{16}.$$
Say $f_1(x) = 1$. If $|x| > 2$ then one of the cases in the table below holds.

shape(x)	$\varepsilon_2(x)$	Lower bound on $g_2(E) - g_1(E)$
$1^1 2^{\frac{n-4}{2}} 3^1$	$\frac{3n-8}{2n(n-1)}$	$\frac{5n-20}{16}$
$1^1 2^{\frac{n-5}{2}} 4^1$	$\frac{n-3}{n(n-1)}$	$\frac{n-3}{4}$
$1^1 2^{\frac{n-7}{2}} 3^2$	$\frac{3n-17}{n(n-1)}$	$\frac{3n-17}{4}$

If $|x| = 2$ then by Lemma 5.0.40(1) we have $\varepsilon_2(x) = 0$, and since n is odd we have
$$g_2(E) - g_1(E) = \frac{n(n-1)}{4}\varepsilon_2(y).$$
Also,
$$o_1(y) = \frac{n-3}{2}$$
and $|y| > 2$. If $f_1(y) \geq 2$ then Lemma 5.0.40(1-6) gives
$$g_2(E) - g_1(E) \geq \frac{n-2}{6}.$$
If $f_1(y) = 1$ then $c_2(y) \geq 2$ and Lemma 5.0.40(7f) gives
$$g_2(E) \geq \frac{n-5}{8}.$$
If $f_1(y) = 0$ then $c_2(y) \geq 2$ and, since n is odd, $|y| \geq 6$. Lemma 5.0.40(8f) gives
$$g_2(E) - g_1(E) \geq \frac{n-3}{8}.$$
Say $f_1(x) = 0$. Then the shape of x is one of $2^{\frac{n-3}{2}} 3^1$ or $2^{\frac{n}{2}}$. In the first case, we have
$$g_2(E) - g_1(E) \geq \frac{n(n-1)}{4}\varepsilon_2(x) = \frac{n-3}{4}.$$
In the second case, we have
$$g_2(E) - g_1(E) = \frac{n(n-1)}{4}\varepsilon_2(y) - \frac{n+2}{8}.$$
Also, $|y| > 2$ and
$$o_1(y) = \frac{n}{2} - 1.$$
Say $f_1(y) \geq 2$. If $|y| = 3$ then $f_1(y) = \frac{n-6}{4}$, so $\varepsilon_2(y) = \frac{n^2-4n-12}{8n(n-1)}$. It now follows from Lemma 5.0.40(2-6) that
$$\varepsilon_2(y) \geq \frac{3n-6}{4n(n-1)}$$

and
$$g_2(E) - g_1(E) \geq \frac{n-10}{16}.$$
If $f_1(y) = 1$ then $c_2(y) \geq 2$ and, since n is even, we have $|y| \geq 6$. Lemma 5.0.40(7f) gives
$$g_2(E) - g_1(E) \geq \frac{n-5}{3} - \frac{n+2}{8} = \frac{5n-46}{24}.$$
If $f_1(y) = 0$ then the shape of y is one of $2^{\frac{n-4}{2}}4^1$ or $2^{\frac{n-6}{2}}3^2$. In the first case, we have
$$\varepsilon_2(y) = \frac{n-6}{2n(n-1)},$$
which gives
$$g_2(E) - g_1(E) = -1.$$
This is impossible by Lemma 2.0.12. In the second case, we have
$$g_2(E) - g_1(E) = \frac{n-8}{2}.$$
Say $g_1(E) = 0$. Equation (5.1) gives
$$g_2(E) = -\frac{n-3}{2} + \frac{n(n-1)}{4}\left(\varepsilon_2(x) + \varepsilon_2(y) + \varepsilon_2(z)\right).$$
Also,
$$o_1(x) + o_1(y) = n + 1.$$
We assume that
$$o_1(x) \geq \frac{n+1}{2},$$
so $o_1(y) \leq \frac{n+1}{2}$ and $f_1(x) > 0$. Also, by Proposition 2.0.16, neither x nor y is homocyclic. In particular,

- $\varepsilon_2(x)$ and $\varepsilon_2(y)$ are nonnegative.

Assume first that $|x| = 2$. Then $|y| > 2$ and $1 \leq f_1(x) \leq n-2$. We use Lemma 5.0.40(1) to determine $\varepsilon_2(x)$. If $f_1(x) = n - 2$ then
$$o_1(y) = 2,$$
so y has shape (a, b). Since $\langle x, y \rangle = S_n$, we have $\gcd(a, b) = 1$ as in Chapter 10. By Theorem 10.0.66(1), we have
$$g_2(E) = 0,$$
and condition 2(c) of the theorem is satisfied.

If $f_1(x) = n - 4$ then we have
$$\varepsilon_2(x) = \frac{2n-10}{n(n-1)},$$
so
$$g_2(E) = \frac{n(n-1)}{4}\varepsilon_2(y) - \begin{cases} 1 & n \text{ odd,} \\ \frac{5}{4} & n \text{ even.} \end{cases}$$
Also,
$$o_1(y) = 3,$$
so y has cycle shape (a, b, c). By Proposition 2.0.16 (applied to y and z),
$$\gcd(a, b, c) = 1.$$

We have
$$\varepsilon_2(y) = \frac{3}{n} - \frac{o_2(y)}{\binom{n}{2}}.$$
We will show that
$$o_2(y) < \begin{cases} \frac{7n-3}{6} & n \text{ odd,} \\ \frac{7n-6}{6} & n \text{ even,} \end{cases}$$
from which it follows that
$$g_2(E) > \frac{n-9}{6}.$$
By Lemma 5.0.38, we have
$$o_2(y) = \sum_{d \in \{a,b,c\}} \lfloor \frac{d}{2} \rfloor + \gcd(a,b) + \gcd(a,c) + \gcd(b,c).$$
At least one of a, b, c is odd, so if n is even then two of a, b, c are odd. Therefore
$$\sum_{d \in \{a,b,c\}} \lfloor \frac{d}{2} \rfloor \leq \begin{cases} \frac{n-1}{2} & n \text{ odd,} \\ \frac{n-2}{2} & n \text{ even.} \end{cases}$$
Note that $c < \frac{n}{3}$. If c divides neither a nor b then
$$\gcd(a,b) + \gcd(a,c) + \gcd(b,c) \leq \frac{n-c}{2} + 2\frac{c}{2} < \frac{2n}{3}$$
and we obtain the claimed bound on $o_2(y)$. If c divides b then $\gcd(a,c) = 1$, so $b \nmid a$. This gives
$$\gcd(a,b) + \gcd(a,c) + \gcd(b,c) \leq \frac{n-c}{4} + 1 + c < \frac{n}{2} + 1$$
and the claimed bound holds. Now assume c divides a. Then $\gcd(b,c) = 1$. If $b \nmid a$ then
$$\gcd(a,b) + \gcd(a,c) + \gcd(b,c) \leq \frac{n-c}{4} + c + 1 < \frac{n}{2} + 1.$$
If $b|a$ then we have $b \leq \frac{a}{2}$ and $c \leq \frac{a}{3}$, so
$$n \leq \frac{11a}{6}$$
and
$$\gcd(a,b) + \gcd(a,c) + \gcd(b,c) = b + c + 1 = n - a + 1 \leq \frac{5n}{11} + 1.$$
In either case the claimed bound holds.

If $6 \leq f_1(x) \leq n-6$ then
$$\varepsilon_2(x) \geq \frac{5n-30}{2n(n-1)},$$
so
$$g_2(E) \geq \frac{n-20}{8}.$$
If $f_1(x) = 5$ then n is odd and
$$\varepsilon_2(x) = \frac{2n-10}{n(n-1)},$$
so
$$g_2(E) = \frac{n(n-1)}{4}\varepsilon_2(y) - 1.$$

If $f_1(y) \geq 2$ then Lemma 5.0.40(2-6) gives
$$g_2(E) \geq \frac{n-8}{6}.$$

We have
$$o_1(y) = \frac{n-3}{2}.$$

Therefore, if $f_1(y) = 1$ then $c_2(y) \geq 2$ and Lemma 5.0.40(7f) gives
$$g_2(E) \geq \frac{n-9}{4}.$$

Also, if $f_1(y) = 0$ then one of the cases in the table below holds.

shape(y)	$\varepsilon_2(y)$	$g_2(E)$
$2^{\frac{n-5}{2}}5^1$	$\frac{2n-10}{n(n-1)}$	$\frac{n-9}{2}$
$2^{\frac{n-7}{2}}3^14^1$	$\frac{2n-10}{n(n-1)}$	$\frac{n-9}{2}$
$2^{\frac{n-9}{2}}3^3$	$\frac{4n-36}{n(n-1)}$	$n-10$

If $f_1(x) = 4$ then n is even and
$$\varepsilon_2(x) = \frac{3n-12}{2n(n-1)},$$
so
$$g_2(E) = \frac{n(n-1)}{4}\varepsilon_2(y) - \frac{n+2}{8}.$$
Also,
$$o_1(y) = \frac{n}{2} - 1.$$

If $f_1(y) \geq 3$ then Lemma 5.0.40(2-5) gives
$$g_2(E) \geq \frac{n-8}{8}.$$

If $f_1(y) = 2$ then $|y| \geq 4$ and Lemma 5.0.40(5) gives
$$g_2(E) \geq \frac{n-10}{16}.$$

If $f_1(y) \leq 1$ then one of the cases in the table below holds.

shape(y)	$\varepsilon_2(y)$	$g_2(E)$
$1^12^{\frac{n-6}{2}}5^1$	$\frac{5n-22}{2n(n-1)}$	$\frac{n-6}{2}$
$1^12^{\frac{n-8}{2}}3^14^1$	$\frac{5n-22}{2n(n-1)}$	$\frac{n-6}{2}$
$1^12^{\frac{n-10}{2}}3^3$	$\frac{9n-78}{2n(n-1)}$	$n-10$
$2^{\frac{n-6}{2}}3^2$	$\frac{5n-30}{2n(n-1)}$	$\frac{n-8}{2}$

If $f_1(x) = 3$ then n is odd and
$$\varepsilon_2(x) = \frac{n-3}{n(n-1)},$$
so
$$g_2(E) = \frac{n(n-1)}{4}\varepsilon_2(y) - \frac{n-3}{4}.$$
Also,
$$o_1(y) = \frac{n-1}{2}.$$

If $|y| = 3$ then
$$f_1(y) = \frac{n-3}{4}.$$

Lemma 5.0.40(2) gives
$$\varepsilon_2(y) = \frac{1}{8}\left(1 - \frac{n+3}{n(n-1)}\right),$$

so
$$g_2(E) = \frac{(n-3)(n-7)}{32}.$$

So we assume that $|y| \geq 4$. If $f_1(y) \geq 3$ then Lemma 5.0.40(3-5) gives
$$g_2(E) \geq \frac{n-3}{8}.$$

If $f_1(y) \leq 2$ then one of the cases in the table below holds.

$shape(y)$	$\varepsilon_2(y)$	$g_2(E)$
$1^2 2^{\frac{n-7}{2}} 5^1$	$\frac{3n-13}{n(n-1)}$	$\frac{n-5}{2}$
$1^2 2^{\frac{n-9}{2}} 3^1 4^1$	$\frac{3n-13}{n(n-1)}$	$\frac{n-5}{2}$
$1^2 2^{\frac{n-11}{2}} 3^3$	$\frac{5n-43}{n(n-1)}$	$n-10$
$1^1 2^{\frac{n-7}{2}} 3^2$	$\frac{3n-17}{n(n-1)}$	$\frac{n-7}{2}$
$1^1 2^{\frac{n-5}{2}} 4^1$	$\frac{n-3}{n(n-1)}$	0
$2^{\frac{n-3}{2}} 3^1$	$\frac{n-3}{n(n-1)}$	0

The last two cases are described in conditions 2(d) and 2(e) of the theorem, respectively, and triples E satisfying these conditions are shown to exist in Proposition 3.0.25.

If $f_1(x) = 2$ then n is even and
$$\varepsilon_2(x) = \frac{n-2}{2n(n-1)},$$

so
$$g_2(E) = \frac{n(n-1)}{4}\varepsilon_2(y) - \frac{3n-8}{8}.$$

Also,
$$o_1(y) = \frac{n}{2}.$$

If $|y| = 3$ then
$$f_1(y) = \frac{n}{4}.$$

Lemma 5.0.40(2) gives
$$g_2(E) = \frac{(n-4)(n-8)}{32}.$$

So we assume that $|y| \geq 4$. If $f_1(y) \geq 4$ then Lemma 5.0.40(3,4) gives
$$g_2(E) \geq \frac{3n-20}{16}.$$

If $f_1(y) \leq 3$ then one of the cases in the table below holds.

shape(y)	$\varepsilon_2(y)$	$g_2(E)$
$1^3 2^{\frac{n-12}{2}} 3^3$	$\frac{11n-96}{2n(n-1)}$	$n-11$
$1^3 2^{\frac{n-10}{2}} 3^1 4^1$	$\frac{7n-32}{2n(n-1)}$	$\frac{n-6}{2}$
$1^3 2^{\frac{n-8}{2}} 5^1$	$\frac{7n-32}{2n(n-1)}$	$\frac{n-6}{2}$
$1^2 2^{\frac{n-8}{2}} 3^2$	$\frac{7n-40}{2n(n-1)}$	$\frac{n-8}{2}$
$1^2 2^{\frac{n-6}{2}} 4^1$	$\frac{3n-8}{2n(n-1)}$	0
$1^1 2^{\frac{n-4}{2}} 3^1$	$\frac{3n-8}{2n(n-1)}$	0

The last two cases are described in conditions 2(f) and 2(g), respectively, and triples E satisfying these conditions are shown to exist in Proposition 3.0.25.

If $f_1(x) = 1$ then n is odd and
$$\varepsilon_2(x) = 0,$$
so
$$g_2(E) = \frac{n(n-1)}{4} \varepsilon_2(y) - \frac{n-3}{2}.$$
Also,
$$o_1(y) = \frac{n+1}{2}.$$
If $|y| = 3$ then
$$f_1(y) = \frac{n+3}{4}.$$
Lemma 5.0.40(2) gives
$$g_2(E) = \frac{(n-5)(n-9)}{32}.$$
So we assume that $|y| \geq 4$. If $f_1(y) \geq 4$ then Lemma 5.0.40(3,4) gives
$$g_2(E) \geq \frac{n-12}{16}.$$
If $f_1(y) \leq 3$ then one of the cases in the following table holds.

shape(y)	$\varepsilon_2(y)$	$g_2(E)$
$1^3 2^{\frac{n-9}{2}} 3^2$	$\frac{4n-24}{n(n-1)}$	$\frac{n-9}{2}$
$1^3 2^{\frac{n-7}{2}} 4^1$	$\frac{2n-6}{n(n-1)}$	0
$1^2 2^{\frac{n-5}{2}} 3^1$	$\frac{2n-6}{n(n-1)}$	0

The second and third cases are described in conditions 2(h) and 2(i), respectively, and triples E satisfying these conditions are shown to exist in Proposition 3.0.25.

Say $|x| = 3$. Since $o_1(x) \geq \frac{n+1}{2}$, we have (under the assumption $n > 20$)
$$f_1(x) \geq \frac{n+3}{4} \geq 6.$$
If $6 \leq f_1(x) \leq n-6$ then Lemma 5.0.40(2) gives
$$\varepsilon_2(x) \geq \frac{4n-24}{n(n-1)}.$$
Therefore
$$g_2(E) \geq \frac{2n-19}{4}.$$

If $f_1(x) = n - 3$ then
$$\varepsilon_2(x) = \frac{2n-6}{n(n-1)}$$
and
$$g_2(E) = \begin{cases} \frac{n(n-1)}{4}\varepsilon_2(y) & n \text{ odd,} \\ \frac{n(n-1)}{4}\varepsilon_2(y) - \frac{1}{4} & n \text{ even.} \end{cases}$$
Also,
$$o_1(y) = 3.$$
We can now argue as we did in the case where x has shape $1^{n-4}2^2$ to obtain
$$g_2(E) \geq \frac{n-3}{6}.$$
We now assume that $|x| \geq 4$. If $f_1(x) \geq 4$ then Lemma 5.0.40(3,4) gives
$$\varepsilon_2(x) \geq \frac{9n-36}{4n(n-1)}.$$
Therefore
$$g_2(E) \geq \frac{n-16}{16}.$$
If $f_1(x) \leq 3$ then one of the cases in the table below holds.

shape(x)	$\varepsilon_2(x)$	$g_2(E) - \frac{n(n-1)}{4}\varepsilon_2(y)$
$1^3 2^{\frac{n-6}{2}} 3^1$	$\frac{5n-18}{2n(n-1)}$	$\frac{n-8}{8}$
$1^3 2^{\frac{n-9}{2}} 3^2$	$\frac{4n-24}{n(n-1)}$	$\frac{n-9}{2}$
$1^3 2^{\frac{n-7}{2}} 4^1$	$\frac{2n-6}{n(n-1)}$	0
$1^2 2^{\frac{n-5}{2}} 3^1$	$\frac{2n-6}{n(n-1)}$	0

In the last two cases, we have $o_1(y) = \frac{n+1}{2}$. If $f_1(y) \geq 2$ then Lemma 5.0.40(2-6) gives
$$g_2(E) \geq \frac{n-2}{8}.$$
If $f_1(y) < 2$ then y has shape $1^1 2^{\frac{n-1}{2}}$ and one of conditions 2(h) or 2(i) holds.

The cases examined above exhaust all possibilities. $\square$

LEMMA 12.0.68. *Let $E = \{x_1, \ldots, x_r\} \subseteq S_n \setminus \{1\}$. Assume that $n > 20$ and that if $r = 3$ then $n > 48$. Assume also that $A_n \leq \langle x_1, \ldots, x_r \rangle$, that $\prod_{i=1}^{r} x_i = 1$ and that x_r has cycle shape (n). Then there is some constant $c' > 0$ such that*
$$g_3(E) - g_2(E) > \max\{\frac{n-10}{3}, c'n^2\}.$$
In particular, $g_3(E) - g_2(E) > 2$.

PROOF. If condition (1) of Theorem 12.0.67 holds then we can prove the lemma in the same manner that we proved Theorem 1.1.2 using Lemma 6.0.55. Thus we may assume that condition (2) of Theorem 12.0.67 holds. Given the results in Chapter 10, we may assume that case 2(c) does not hold.

In Chapter 10 we showed that if z is an n-cycle then
$$i_3(z) = \begin{cases} \binom{n}{3} - \frac{n^2-3n+6}{6} & 3 | n, \\ \binom{n}{3} - \frac{(n-1)(n-2)}{6} & 3 \nmid n. \end{cases}$$

For all other x in all E under consideration, every nontrivial power of x has one of the shapes given in the table below.

$shape(y)$	$f_3(y)$
$1^a 2^b$	$\binom{a}{3} + ab$
$1^a 2^b 3^1$	$\binom{a}{3} + ab + 1$
$1^a 2^b 4^1$	$\binom{a}{3} + ab$

Using Burnside's lemma and the results in the table above, we can calculate $i_3(x)$ for all remaining $x \in E$. We obtain the following results for $g_3(E)$.

Condition satisfied	$g_3(E)$ $(3\mid n)$	$g_3(E)$ $(3 \nmid n)$
$2(a),(b),(e),(g),(i)$	$\frac{(n-3)(n-6)}{6}$	$\frac{(n-4)(n-5)}{6}$
$2(d)$	$\frac{n^2-12n+33}{6}$	$\frac{(n-5)(n-7)}{6}$
$2(f)$	$\frac{(n-6)^2}{6}$	$\frac{n^2-12n+38}{6}$
$2(h)$	$\frac{n^2-12n+39}{6}$	$\frac{n^2-12n+41}{6}$

□

The proof of Theorem 1.2.1 is now completed in the same manner as that of Theorem 1.1.2. Using Lemma 2.0.13, Theorem 12.0.67 and Lemma 12.0.68, we see that of (G, H, E) is a counterexample to the Theorem with $n > 48$ then either H is a primitive 3-homogeneous subgroup of G or E is as described in of the conditions (f),(g) in the theorem and H is the stabilizer of a partition of $[n]$ into two parts of size $n/2$ or $n/2$ parts of size two. In the primitive 3-homogeneous case, the theorem (for $n > 48$) now follows from Theorem 8.0.63. Now we calculate the following results.

Condition	$n \bmod 4$	$g_4(E)$
(f)	0	$\frac{(n-4)^2(n-7)}{16}$
(f)	2	$\frac{(n-3)(n-6)^2}{16}$
(g)	0	$\frac{(n-4)(n^2-7n+4)}{16}$
(g)	2	$\frac{(n-6)(n^2-5n+2)}{16}$

It follows that $g_4(E) - g_3(E) \geq 7$ whenever $n \geq 10$, and we can apply Lemma 2.0.13.

When $r \geq 5$ the Theorem follows from Theorem 1.1.2, so it remains to examine the finitely many cases where $n \leq 48$ and $r \leq 4$. These cases are handled by direct computer calculations, as described in Appendix A, along with Theorem 8.0.63.

CHAPTER 13

Galois groups of trinomials - the proofs of Propositions 1.4.1 and 1.4.2 and Theorem 1.4.3

In this chapter we prove the theorems stated in Chapter 1.4, which we summarize in the theorem below. Recall from Chapter 1.4 that for any positive integer n, the set $\mathcal{S}'(n)$ consists of S_n, A_n, and all subgroups of S_n which are the full stabilizer in S_n of a point, two points or a 2-set. For $a \in [n]$ with $\gcd(n,a) = 1$, the set $\mathcal{S}(n,a)$ consists of all subgroups $H \leq S_n$ such that there exist a Riemann surface $X = X(H,a)$ with $g(X) \leq 2$ and an analytic map $f : X \to \mathbf{P}^1$ with monodromy data $(S_n, H, E = (t_a, z, y_a))$, where $t_a = (1, a+1)$, $z = (1, 2, \ldots, n)$ and $t_a z y_a = 1$.

THEOREM 13.0.69. (1) *(Proposition 1.4.1)* For all n, a we have $\mathcal{S}'(n) \subseteq \mathcal{S}(n,a)$.

(2) *(Proposition 1.4.2)* If $n \leq 4$ each $\mathcal{S}(n,a)$ contains all subgroups of S_n.

(3) *(Theorem 1.4.3)* If $n > 10$ then $\mathcal{S}(n,a) = \mathcal{S}'(n)$, and if $5 \leq n \leq 10$ then $\mathcal{S}(n,a) \setminus \mathcal{S}'(n)$ consists of those groups listed in the tables given in Theorem 1.4.3.

Given $H \leq S_n$ and $a \in [n]$ with $\gcd(n,a) = 1$, let $X(H,a)$ be any Riemann surface such that there is an analytic map from $X(H,a)$ to $\mathbf{P}^1$ with monodromy data $(S_n, H, (t_a, z, y_a))$. We begin by recording the following facts, which follow immediately upon combining Lemma 2.0.11 with basic facts about induced modules (see [**Is**]). These facts may be used without reference to the given proposition in what follows.

PROPOSITION 13.0.70. (1) If $H, K \leq S_n$ are conjugate then $g(X(H,a)) = g(X(K,a))$ for all a.
(2) If $H \leq K \leq S_n$ then $g(X(H,a)) \geq g(X(K,a))$ for all a.

Proposition 1.4.2 now follows from easy computations with $H = 1$.

Next we prove Proposition 1.4.1. Since every element of $\mathcal{S}'(n)$ contains either some two point stabilizer $H_{i,j}$ or A_n, it suffices to show that $g(X) = 0$ when H is one of $H_{1,2}$ or A_n. First assume that $H = H_{1,2}$, and for $g \in S_n$ let $o(g)$ be the number of orbits of $\langle g \rangle$ on ordered pairs of distinct elements of $[n]$. The Riemann-Hurwitz formula gives

$$2(g(X) - 1) = n(n-1) - o(t_a) - o(z) - o(y_a).$$

Using Burnside's Lemma, we get

$$o(t_a) = \frac{1}{2}[n(n-1) + (n-2)(n-3)] = n^2 - 3n + 3,$$

and

$$o(z) = \frac{1}{n}n(n-1) = n - 1.$$

Let σ_1, σ_2 be the cycles of y_a, with $|supp(\sigma_1)| = a$, $|supp(\sigma_2)| = n - a$. Since $\gcd(a, n - a) = 1$, we see that $\langle y_a \rangle$ is transitive on $supp(\sigma_1) \times supp(\sigma_2)$ and on $supp(\sigma_2) \times supp(\sigma_1)$. Also, using Burnside's Lemma as we did for z, we see that $\langle y_a \rangle$ has $a - 1$ orbits on pairs from $supp(\sigma_1)$ and $n - a - 1$ orbits on pairs from $supp(\sigma_2)$. Thus

$$o(y_a) = n,$$

and direct calculation gives $g(X) = 0$. A similar but easier argument shows that $g(X) = 0$ when $H = A_n$, as each $g \in S_n \setminus A_n$ has one orbit on the cosets of A_n while each $g \in A_n$ has two such orbits. Note that for any n we have $z \in A_n$ if and only if $y_a \notin A_n$.

It remains to prove Theorem 1.4.3. Certainly $\mathcal{S}(n, a) = \mathcal{S}(n, n - a)$, so we assume from now on that $a > \frac{n}{2}$.

Assume until otherwise indicated that $n > 10$. By Theorem 10.0.66 we have $g_3(E) - g_2(E) > 2$ and $g_4(E) - g_3(E) > 2$. Combining this fact with Lemma 2.0.13 gives the following key result.

LEMMA 13.0.71. *If $n > 10$ and $g(X(H, a)) \leq 2$ then $o_2(H) = o_3(H) = o_4(H)$.*

We proceed by examining first the case where H is transitive and then the case where H is intransitive. So, assume that H is transitive. If H, K are subgroups of S_n with $H \leq K$ then $g(X(H)) \geq g(X(K))$. So, we begin by examining each maximal transitive proper subgroup of S_n other than A_n. If such a subgroup is not primitive then it is the full stabilizer in S_n of a partition of $[n]$ into $\frac{n}{d}$ parts of size d for some nontrivial proper divisor d of n. We saw while proving Theorem 1.1.2 in Chapter 4 that for any such subgroup H, either $o_3(H) > o_2(H)$ or $o_4(H) > o_3(H)$. Thus $g(X(H)) > 2$ by Lemma 13.0.71

Now assume that H is primitive. It follows from Lemma 4.0.32 that if $g(X) \leq 2$ then $o_4(H) = 1$, that is, H is 4-homogeneous. We now invoke the following results. The first is due to W. Kantor (see [**Ka**]) and the second follows from the classification of finite simple groups (see for example [**Ca**]).

LEMMA 13.0.72. *If $H \leq S_n$ is 4-homogeneous but not 4-transitive then*
- *$n = 9$ and $L_2(8) \leq H \leq Aut(L_2(8))$, or*
- *$n = 33$ and $H = Aut(L_2(32))$.*

LEMMA 13.0.73. *If $H \leq S_n$ is 4-transitive and $A_n \not\leq H$ then $n \in \{11, 12, 23, 24\}$ and H is the Mathieu group M_n.*

As we are assuming that $n \geq 10$, we must examine $Aut(L_2(32))$ in its action on 1-spaces from the natural vector space and the Mathieu groups in their natural actions. Let H be one of these groups and let Ω be the set of right cosets of H in S_n. For $y \in E$, let $o_H(y)$ be the number of orbits of y on Ω. The Riemann-Hurwitz formula gives

$$2(g(X) - 1) = |\Omega| - \sum_{y \in E} o_H(y).$$

Since no primitive proper subgroup of S_n contains a transposition, t_a fixes no element of Ω, so

$$o_H(t_a) = \frac{|\Omega|}{2}$$

by Burnside's lemma. For $y \in S_n$, let $f_H(y)$ be the number of cosets of H fixed by y, and let $Cl(y)$ be the conjugacy class of y. Then

$$f_H(y) = \frac{|C_{S_n}(y)||Cl(y) \cap H|}{|H|} = [S_n : H]m(S_n, H),$$

where $m(G, H)$ is as defined in Chapter 8

Define

$$M := M(H) := \max\{f_H(y) : y \in S_n \setminus \{1\}\}.$$

Using Burnside's Lemma, we see that for any primitive $H \leq S_n$ we have

$$\begin{aligned} 2(g(X) - 1) &= \frac{|\Omega|}{2} - o_H(y_a) - o_H(z) \\ &= \left(\frac{1}{2} - \frac{1}{a(n-a)} - \frac{1}{n}\right)|\Omega| \\ &\quad - \frac{1}{a(n-a)}\sum_{j=1}^{a(n-a)-1} f_H(y_a^j) - \frac{1}{n}\sum_{j=1}^{n-1} f_H(z^j) \\ &\geq \left(\frac{1}{2} - \frac{1}{a(n-a)} - \frac{1}{n}\right)|\Omega| - \left(\frac{a(n-a) - 1}{a(n-a)} + \frac{n-1}{n}\right)M \\ &= \frac{1}{2}|\Omega| - 2M - \left(\frac{1}{a(n-a)} + \frac{1}{n}\right)(|\Omega| - M) \\ &\geq \frac{1}{2}|\Omega| - 2M - \frac{2n-1}{n(n-1)}(|\Omega| - M). \end{aligned}$$

Using [**CCNPW**], we get the results in the table below.

n	H	$M(H)$	Ω
11	M_{11}	48	7!
12	M_{12}	192	7!
23	M_{23}	19,353,600	19!/48
24	M_{24}	255,467,520	19!/48
33	$Aut(L_2(32))$	$2^{11}16!/5$	30!/5

Easy calculations show that $g(X) > 2$ in each of these cases.

Assume now that H is intransitive. Let $\lambda(H) = (\lambda_1, \ldots, \lambda_r)$ be the partition of n determined by the sizes of the orbits of H on $[n]$. If $\lambda_1 < n - 2$ then H fixes some k-subset of $[n]$ with $3 \leq k \leq n - 3$. This gives

$$g(X) \geq g_k(E) \geq g_3(E) > 2,$$

the last inequality following from Theorem 10.0.66, given our assumption that $n > 10$. So, we may assume that $\lambda_1 \in \{n-1, n-2\}$.

First assume that $\lambda_1 = n - 1$. We may (and do) assume that $[n-1]$ is an orbit of H. For $1 \leq k \leq n-1$, let $o'_k(H)$ be the number of orbits of H on k-subsets of $[n-1]$. Then for $k \geq 2$ we have

$$o_k(H) = o'_k(H) + o'_{k-1}(H).$$

This gives

$$o_3(H) - o_2(H) = o'_3(H) - 1.$$

It follows that if $g(X) \leq 2$ then $o'_3(H) = 1$, so H acts primitively on $[n-1]$ and $o'_2(H) = 1$. It then follows that
$$o_4(H) - o_3(H) = o'_4(H) - 1,$$
so if $g(X) \leq 2$ then $o'_4(H) = 1$. By Lemmas 13.0.72 and 13.0.73, we see that one of the following cases holds.
  (1) H acts as S_{n-1} on $[n-1]$.
  (2) H acts as A_{n-1} on $[n-1]$.
  (3) H acts as the Mathieu group M_{n-1} on $[n-1]$ and $n \in \{12, 13, 24, 25\}$.
  (4) H acts on $[n-1]$ as $Aut(L_2(32))$ acts on 1-spaces from its natural vector space, and $n = 34$.

If the first case listed above holds then $H = H_n$ and $H \in \mathcal{S}'(n)$. If the second case holds then $H = H_n \cap A_n$. If either of the last two cases holds then H is contained in the group $H_n \cap A_n$ described in the second case. Thus it suffices for our purposes to show that $g(X) > 2$ in the second case. In that case, H is the stabilizer of a point in the action of S_n on $[n] \times \{1, -1\}$ defined by
$$(i, j)\sigma := (i\sigma, sign(\sigma)j).$$
This allows us to calculate easily that $o_H(t_a) = n$, $o_H(z) = 2$ and
$$o_H(y_a) = \begin{cases} 4 & n \text{ even,} \\ 3 & n \text{ odd.} \end{cases}$$
The Riemann-Hurwitz formula now gives
$$g(X) = \begin{cases} \frac{n-4}{2}, & n \text{ even,} \\ \frac{n-3}{2}, & n \text{ odd.} \end{cases}$$
Thus $g(X) > 2$ whenever $n \geq 9$.

Now assume that $\lambda_1 = n - 2$. Note that if $\lambda(H) = (n-2, 1, 1)$ then there is some K such that $H < K \leq S_n$ and $\lambda(K) = (n-2, 2)$. Namely, we can take $K = H \times \langle x \rangle$, where x is the transposition whose support consists of the two fixed points of H. Note that $K = H_{\{i,j\}}$ if and only if $H = H_{i,j}$. It therefore suffices for our purposes to show that if $\lambda(H) = (n-2, 2)$ and $g(X) \leq 2$ then $H = H_{\{i,j\}}$ for some 2-set $\{i, j\} \subseteq [n]$. We may assume that the orbits of H are $[n-2]$ and $\{n-1, n\}$, so $H \leq H_{\{n-1,n\}}$. Let $P = H_{n-1,n}$ and let Q be generated by the transposition $(n-1, n)$, so $H_{\{n-1,n\}} = P \times Q$. Let $\pi(H)$ be the projection of H onto P.

We show first that if $g(X) \leq 2$ then $\pi(H) = P$. Assume for contradiction that $\pi(H) \neq P$. Since $H \leq \pi(H) \times Q$, it suffices to consider the case where $H = \pi(H) \times Q$. For $1 \leq k \leq n-2$ let $p_k(H)$ be the number of orbits of $\pi(H)$ on k-subsets of $[n-2]$. Then
$$o_2(H) = p_2(H) + p_1(H) + 1 = p_2(H) + 2,$$
$$o_3(H) = p_3(H) + p_2(H) + p_1(H) = p_3(H) + p_2(H) + 1$$
and
$$o_4(H) = p_4(H) + p_3(H) + p_2(H).$$
This gives
$$o_3(H) - o_2(H) = p_3(H) - 1$$
and
$$o_4(H) - o_3(H) = p_4(H) - 1.$$

It follows as it did in the case $\lambda_1 = n-1$ that if $g(X) \leq 2$ then P is 4-homogeneous on $[n-2]$, and that it suffices for our current purposes to show that $g(X) > 2$ when $P = A_{n-2}$. Let Γ be the set of ordered pairs of distinct elements of $[n]$. Then S_n acts transitively on $\Gamma \times \{1, -1\}$ by

$$((a,b), j)\sigma = ((a\sigma, b\sigma), sign(\sigma)j).$$

Define an equivalence relation $\sim$ on $\Gamma \times \{1, -1\}$ by

$$((a,b), j) \sim ((b,a), -j).$$

This relation is S_n-invariant and if $P = A_{n-2}$ then H is the stabilizer of an equivalence class. It is now straightforward to compute that

$$o_H(t_a) = \binom{n}{2} + 1,$$

$$o_H(z) = \begin{cases} n & n \equiv 2 \bmod 4, \\ n-1 & \text{otherwise,} \end{cases}$$

and

$$o_H(y_a) = \begin{cases} n & n \text{ even,} \\ \frac{2n-a+1}{2} & n \text{ odd}, n-a \equiv 0 \bmod 4, \\ \frac{2n-a+3}{2} & n \text{ odd}, n-a \equiv 2 \bmod 4, \\ \frac{n+a+1}{2} & n \text{ odd}, a \equiv 0 \bmod 4, \\ \frac{n+a+3}{2} & n \text{ odd}, a \equiv 2 \bmod 4. \end{cases}$$

This gives

$$g(X) = \begin{cases} \frac{n^2-5n+4}{4} & n \equiv 0 \bmod 4, \\ \frac{n^2-5n+2}{4} & n \equiv 2 \bmod 4, \\ \frac{n^2-5n+a+3}{4} & n \text{ odd}, n-a \equiv 0 \bmod 4, \\ \frac{n^2-5n+a+1}{4} & n \text{ odd}, n-a \equiv 2 \bmod 4, \\ \frac{n^2-4n-a+3}{4} & n \text{ odd}, a \equiv 0 \bmod 4, \\ \frac{n^2-4n-a+1}{4} & n \text{ odd}, a \equiv 2 \bmod 4, \end{cases}$$

and it follows that $g(X) > 2$ whenever $n > 6$.

So we must now examine the case where $\pi(H) = P$. Using well known results about subgroups of direct products and the normal subgroups of S_n, we see that in this case we have either $H = P \times Q = H_{\{n-1,n\}}$ or $H = (P \times Q) \cap A_n$. It remains to show that in the second case we have $g(X) > 2$, so assume this second case holds. Let Λ be the set of 2-subsets of $[n]$. Then S_n acts transitively on $\Lambda \times \{1, -1\}$ by

$$(\{a,b\}, j)\sigma = (\{a\sigma, b\sigma\}, sign(\sigma)j),$$

and H is the stabilizer of a point in this action. Using this fact, it is straightforward to compute that

$$o_H(t_a) = \binom{n}{2},$$

$$o_H(z) = \begin{cases} n & 4|n, \\ n-1 & \text{otherwise,} \end{cases}$$

and
$$o_H(y_a) = \begin{cases} n & n \text{ even,} \\ \frac{2n-a+3}{2} & n \text{ odd}, n-a \equiv 0 \bmod 4, \\ \frac{2n-a+1}{2} & n \text{ odd}, n-a \equiv 2 \bmod 4, \\ \frac{n+a+3}{2} & n \text{ odd}, a \equiv 0 \bmod 4, \\ \frac{n+a+1}{2} & n \text{ odd}, a \equiv 2 \bmod 4. \end{cases}$$

This gives
$$g(X) = \begin{cases} \frac{n^2-5n+4}{4} & n \equiv 0 \bmod 4, \\ \frac{n^2-5n+6}{4} & n \equiv 2 \bmod 4, \\ \frac{n^2-5n+a+3}{4} & n \text{ odd}, n-a \equiv 0 \bmod 4, \\ \frac{n^2-5n+a+5}{4} & n \text{ odd}, n-a \equiv 2 \bmod 4, \\ \frac{n^2-4n-a+3}{4} & n \text{ odd}, a \equiv 0 \bmod 4, \\ \frac{n^2-4n-a+5}{4} & n \text{ odd}, a \equiv 2 \bmod 4. \end{cases}$$

It follows that $g(X) > 2$ whenever $n > 5$.

The proof that $\mathcal{S}(n,a) = \mathcal{S}'(n)$ whenever $n > 10$ is now complete. It remains to examine the case $n \leq 10$. In this case the subgroup lattice of S_n is well understood. For each a we can find all $H \leq S_n$ such that $g(X(H,a)) \leq 2$ using the following algorithm.

(1) Let $\mathcal{M}_0 = \mathcal{S}'(n)$
(2) For $i > 0$, let $\mathcal{M}_i$ be the set of all $K < S_n$ such that $K \notin \bigcup_{j<i} \mathcal{M}_j$, but every $H \leq S_n$ which properly contains K lies in $\bigcup_{j<i} \mathcal{M}_j$ and satisfies $g(X(H,a)) \leq 2$.
(3) Determine $g(X(K,a))$ for all $K \in \mathcal{M}_i$.
(4) If there is some $K \in \mathcal{M}_i$ such that $g(X(K)) \leq 2$ then increase i by one and return to step 2. Otherwise, stop.

Upon carrying out this algorithm for each n between 5 and 10 and each $a \in [n]$ such that $\gcd(n,a) = 1$, we see that the only elements of $\mathcal{S}(n,a) \setminus \mathcal{S}'(n)$ are those described in Tables 1 and 2 of Theorem 1.4.3.

APPENDIX A

Finding small genus examples by computer search - by R. Guralnick and R. Stafford

We assume in all cases that $G = A_n$ or S_n with $n \geq 5$.

A.1. Description

We use MAGMA to do the necessary computations.

We want to describe the instances when $g_2(E)$ and $g_3(E)$ are small for generating systems E of S_n and A_n. Both for computational reasons and for considering other actions, it is more useful to consider the differences, $g_2(E) - g_1(E)$ and $g_3(E) - g_2(E)$. As shown in the paper (see Lemma 2.0.12), these differences are always nonnegative.

The idea we use is to loop through representatives for all the conjugacy classes of S_n and compute the differences $i_2(x)/\binom{n}{2} - i_1(x)/n$ and $i_3(x)/\binom{n}{3} - i_2(x)/\binom{n}{2}$. We then group all the classes which have these same differences. There are not very many classes with negative differences and indeed most of the time the differences are positive. On the other hand, unless $g_1(E) \leq 1$, some negative terms will need to be present and large positive terms cannot occur for a fixed number branch of points r (particularly for $r = 3, 4$). This allows us to find all possible E with $g_k(E)$ small for n up to 57 (note that S_{57} has quite a large number of conjugacy classes).

We compute all possible sets of conjugacy classes which satisfy these numerical conditions. We then eliminate many of the possible by E using the fact that $g_1(E) \geq 0$. We further eliminate many possibilities by noting that if $g_1(E) = 0$, then the greatest common denominator of the cycle lengths of any two of the generators must be relatively prime (see Proposition 2.0.16).

We also send to a separate file solutions that we know exist. One thing that we have not done is to determine whether all these systems do exist. For small n, this can be done fairly easily. However, for larger n, particularly for $r > 3$, this can be a complicated calculation.

A.2. $n = 5$ and $n = 6$

If $n \leq 6$, there are too many small genus systems to enumerate explicitly for $r = 5$.

THEOREM A.2.1. *Assume that $r \geq 6$, $n = 5$ or 6 and $g_2(E) \leq 2$. Then one of the following holds:*

(1) $r = 7$, $n = 5$, E *contains six transpositions and one element of cycle shape* $1^1 2^2$, *and* $g_2(E) = 2$;
(2) $r = 7$, $n = 6$, E *contains five transpositions, one fixed point free involution and one element of cycle shape* $1^2 2^2$, *and* $g_2(E) = 2$;

(3) $r = 6$ and $g_2(E) > 0$.

If we assume that one of the generators is an n-cycle, then there are no exceptions for $r \geq 5$.

THEOREM A.2.2. *Assume that $r \geq 5$, $n = 5$ or 6 and E contains an n-cycle. Then $g_2(E) > 2$.*

If $n = 5$, then aside from the action on k-sets for $k \leq 2$, there is only one other primitive action. It has degree 6. Since no element acts as a transposition in this case, we see that if E has r terms, then $g(G, H, E) \geq (r - 5)$.

If $n = 6$, then the primitive actions of G of cardinality greater than 6 are either on 2-sets or acting by conjugation on the set of fixed point free involutions. The stabilizer of a 2-set is the centralizer of transposition.

A.3. $5 \leq r \leq 8$, $7 \leq n \leq 20$

THEOREM A.3.1. *Assume that $r \geq 5$ and $7 \leq n \leq 20$. Then $g_3(E) - g_2(E) > 2$ and $g_3(E) > 5$ unless $n = 7$, $r = 5$ and one of the following holds:*
  (1) *E contains three elements of cycle shape $1^1 2^3$, one of shape $1^3 2^2$ and one of shape $1^5 2^1$, in which case $g_1(E) = g_2(E) = 0$ and $g_3(E) = 2$; or*
  (2) *$g_3(E) - g_2(E) = 2$ and $g_3(E) \geq 3$.*

THEOREM A.3.2. *Assume that $r \geq 5$ and $7 \leq n \leq 20$. Then one of the following occurs:*
  (1) *$g_2(E) > 2$, $g_3(E) > 5$ and $g_2(E) - g_1(E) > 2$; or*
  (2) *$r = 5$, E is as described in Proposition 3.0.24(a,b), $g_2(E) = g_1(E) = 0$ and $g_3(E) \geq 6$;*
  (3) *$r = 6$, $n = 7$ and E consists of three involutions with one fixed point and three transpositions and $(g_1(E), g_2(E), g_3(E)) = (0, 1, 5)$;*
  (4) *$r = 6$, $n = 7$ and E consists of two involutions in each of the three conjugacy classes of involutions and $(g_1(E), g_2(E), g_3(E)) = (0, 2, 6)$;*
  (5) *$r = 6$, $n = 7$ and $(g_1(E), g_2(E), g_3(E)) = (1, 3, 8)$;*
  (6) *$r = 6$, $n = 8$ and E consists of involutions and $(g_1(E), g_2(E), g_3(E)) = (0, 2, 10)$;*
  (7) *$r = 6$, $n = 8$ and E consists of involutions and $(g_1(E), g_2(E), g_3(E)) = (1, 2, 13)$; or*
  (8) *$r = 6$, $n = 8$, and $(g_1(E), g_2(E)) = (1, 3)$ or $(2, 3)$ with $g_3(E) \geq 13$;*
  (9) *$r = 6, n = 9$ and $(g_1(E), g_2(E), g_3(E)) = (0, 2, 14)$;*
  (10) *$r = 6$, $n = 10$ and $(g_1(E), g_2(E), g_3(E)) = (1, 3, 25)$ or $(2, 4, 29)$;*
  (11) *$r = 5$, $n = 7$ and $(g_1(E), g_2(E), g_3(E))$ is one of $(0, 1, \geq 3)$, $(0, 2, \geq 4)$, $(1, 2, \geq 5)$, or $(1, 3, \geq 6)$;*
  (12) *$r = 5$, $n = 8$ and (g_1, g_2, g_3) is one of $(0, 1, \geq 6)$, $(0, 2, \geq 10)$, $(1, 1, 9)$, $(1, 2, \geq 9)$, $(1, 3, \geq 11)$, $(2, 2, 12)$, $(2, 3, \geq 12)$ or $(2, 4, \geq 14)$;*
  (13) *$r = 5$, $n = 9$ and $(g_1(E), g_2(E), g_3(E))$ is one of $(0, 1, 9)$, $(0, 2 \geq 10)$, or $(1, 3, 13)$;*
  (14) *$r = 5$, $n = 10$, and $(g_1(E), g_2(E), g_3(E))$ is one of $(0, 2, \geq 14)$, $(1, 2, 17)$, $(1, 3, \geq 19)$, $(2, 3, \geq 23)$, or $(2, 4, \geq 26)$.*
  (15) *$r = 5$, $n = 11$, and $(g_1(E), g_2(E), g_3(E)) = (0, 2, \geq 18)$;*
  (16) *$r = 5$, $n = 12$ and $(g_1(E), g_2(E), g_3(E))$ is one of $(0, 2, 26)$, $(1, 2, 27)$, $(1, 3, \geq 31)$, or $(2, 4, 38)$;*

(17) $r = 5$, $n = 13$, $(g_1(E), g_2(E), g_3(E)) = (0, 2, 28)$;
(18) $r = 5$, $n = 14$, $(g_1(E), g_2(E), g_3(E)) = (1, 3, 41)$; or
(19) $r = 5$, $n = 16$, and $(g_1(E), g_2(E), g_3(E)) = (1, 3, 57)$.

COROLLARY A.3.3. *If $r \geq 5$, $7 \leq n \leq 20$ and H is a maximal subgroup other than the stabilizer of a k-set for $k < 3$, then $g(G, H, E) > 2$.*

PROOF. For k-sets, this follows from the previous result and the fact that $g_k(E) \geq g_{k-1}(E)$. If H is transitive and not 2-homogeneous, then $(1_H^G, 1_{G_2}^G - 1_{G_1}^G) > 0$, whence $g(G, H, E) > g_2(E) - g_1(E)$. By the previous result, this is at least 3 except for a few specific choices for E. We compute those directly and verify the result.

If H is 2-homogeneous, but not 3-homogeneous, then we use the same argument to note that $g(G, H, E) \geq g_3(E) - g_2(E)$ and we apply Theorem A.3.1 to conclude the result (noting that when $n = 7$, there are no three homogeneous subgroups). If H is 3-homogeneous, we apply the results of §8. □

A.4. $r < 5$

We now consider cases with $r \in \{3, 4\}$. We only consider systems where E contains an n-cycle (if $r = 4$, these always lead to systems with $r = 5$ not necessarily containing an n-cycle with g_1 and g_2 unchanged).

THEOREM A.4.1. *Say $r \leq 4$, $7 \leq n \leq 20$ and E contains an n-cycle. Then one of the following holds.*

(1) *E is given in Theorem 1.2.1 and $g_2(E) = g_1(E) = 0$.*
(2) *We have $g_3(E) - g_2(E) > 2$.*
(3) *We have $r = 3$ and $n \leq 9$.*
(4) *E consists of one element of cycle shape $1^4 2^3$, one of shape $1^1 3^3$ and one of shape 10^1, and $(g_1(E), g_2(E), g_3(E)) = (0, 0, 0)$.*
(5) *E consists of one element of cycle shape $1^3 2^4$, one of shape $1^2 3^3$ and one of shape 11^1, and $(g_1(E), g_2(E), g_3(E)) = (0, 1, 3)$.*
(6) *E consist of one element of cycle shape $1^5 2^1$, one of shape $1^3 2^2$, one of shape $1^1 2^3$ and one of shape 7^1, and $(g_1(E), g_2(E), g_3(E)) = (0, 0, 1)$.*
(7) *E consists of two elements of cycle shape $1^5 2^1$, one of shape $1^1 3^2$ and one of shape 7^1, and $(g_1(E), g_2(E), g_3(E)) = (0, 1, 2)$.*
(8) *E consists of one element of cycle shape $1^6 2^1$, two of shape $1^2 2^3$ and one of shape 8^1, and $(g_1(E), g_2(E), g_3(E)) = (0, 0, 2)$.*

THEOREM A.4.2. *If $r \leq 4$, E contains an n-cycle and $20 < n < 57$ then $g_3(E) - g_2(E) > 2$.*

There are too many examples with $g_2 = 1$ or 2 to list. We do have:

THEOREM A.4.3. *If $r \leq 4$, E contains an n-cycle and $g_2 \leq 2$, then $n \leq 16$ or E is given in Theorem 1.2.1.*

If we consider other actions, we have the following result.

THEOREM A.4.4. *If $H \neq A_n$ is a maximal subgroup of G other than the stabilizer of a k-set for $k \leq 2$, $r \leq 4$, E contains an n-cycle and $7 \leq n \leq 57$, then $g(G, H, E) > 0$ unless $n = 8, G = S_8$, $H = S_4 \wr S_2$, $|E| = 3$ and consists of an 8-cycle, and either an involution with 2 fixed points and an element of shape $4^1 2^1 1^2$ or an involution with 4 fixed points and an element of shape $3^2 2^1$.*

PROOF. We note that if $g_2 = g_1$, then either E is one of the standard examples or $r = 3$ and $n \leq 11$. Moreover, if $g_3 > g_2$ even in those cases, the result holds unless H is 3-homogeneous, a case already handled or n is even and H is imprimitive. In the remaining cases, we just compute. □

If $n = 8$, there are other examples of genus 1 or 2. We do not list these here.

Bibliography

[ArPi] M. Artebani and G. Pirola, Algebraic functions with even monodromy, *Proc. Amer. Math. Soc.* **133** (2005), 331–341.

[As] M. Aschbacher, On conjectures of Guralnick and Thompson, *J. Algebra* **135** (1990), no. 2, 277-343.

[Bo] B. Bollobás, *Combinatorics. Set systems, hypergraphs, families of vectors and combinatorial probability*, Cambridge University Press, Cambridge-New York, 1986.

[Ca] P. J. Cameron, Permutation Groups, *Handbook of Combinatorics, Vol. 1*, R. L. Graham et. al., editors, Elsevier, Amsterdam, 1995.

[CaNeSa] P. J. Cameron, P. M. Neumann and J. Saxl, An interchange property in finite permutation groups, *Bull. London Math. Soc.* **11** (1979), 161-169.

[CCNPW] J. H. Conway, R. T. Curtis, S. P. Norton, R. A. Parker, R. A. Wilson, *Atlas of finite groups. Maximal subgroups and ordinary characters for simple groups. With computational assistance from J. G. Thackray*, Oxford University Press, Oxford, 1985.

[DiMo] J. D. Dixon and B. Mortimer, *Permutation Groups*, Springer-Verlag, New York, 1996.

[El] N. Elkies, personal communication, 1999.

[Fa] G. Faltings, Endlichkeitssätze für abelsche Varietäten ber Zahlkörpern, *Invent. Math.* **73** (1983), 349-366.

[FeLySc] W. Feit, R. Lyndon and L. L. Scott, A remark about permutations, *J. Combinatorial Theory Ser. A* **18** (1975), 234-335.

[Fri1] M. Fried, Galois groups and complex multiplication, *Trans. Amer. Math. Soc.* **237** (1978), 141-162.

[Fri2] M. Fried, Combinatorial computation of moduli dimension of Nielsen classes of covers, *Graphs and algorithms (Boulder, CO, 1987)*, 61-79, Contemp. Math., **89**, Amer. Math. Soc., Providence, RI, 1989.

[FriGu] M. Fried and R. M. Guralnick, On uniformization of generic curves of genus g by radicals, preprint.

[FGM] D. Frohardt, R. Guralnick and K. Magaard, Genus 0 actions of groups of Lie rank 1, *Arithmetic fundamental groups and noncommutative algebra (Berkeley, CA, 1999)*, 449-483, Proc. Sympos. Pure Math., **70**, Amer. Math. Soc., Providence, RI, 2002.

[FrMa] D. Frohardt and K. Magaard, Composition factors of monodromy groups, *Ann. of Math.* (2) **154** (2001), no. 2, 327-345.

[Gu1] R. M. Guralnick, The genus of a permutation group, *Groups, combinatorics & geometry (Durham, 1990)*, 351-363, London Math. Soc. Lecture Note Ser., **165**, Cambridge Univ. Press, Cambridge, 1992.

[Gu2] R. M. Guralnick, Monodromy groups of covers of curves, *Galois Groups and Fundamental Groups*, Edited by Leila Schneps, 1-46, Math. Sci. Res. Inst. Publ. **41**, Cambridge University Press, Cambridge, 2003.

[GuMa] R. M. Guralnick and K. Magaard, On the minimal degree of a primitive permutation group, *J. Algebra* **207** (1998), no. 1, 127-145.

[GuNe] R. M. Guralnick and M. Neubauer, Monodromy groups of branched coverings, *Recent Developments in the Inverse Galois Problem*, M. Fried et. al., editors, American Mathematical Society, Providence, RI, 1995.

[GuTh] R. M. Guralnick and J. Thompson, Finite groups of genus zero, *J. Algebra* **131** (1990), 303-341.

[Is] I. M. Isaacs, *Character Theory of Finite Groups*, Dover, New York, 1994.

[JaKe]   G. James and A. Kerber, *The Representation Theory of the Symmetric Group*, With a foreword by P. M. Cohn. With an introduction by Gilbert de B. Robinson. Encyclopedia of Mathematics and its Applications, 16, Addison-Wesley Publishing Co., Reading, Mass., 1981.

[Ka]   W. Kantor, k-homogeneous groups, *Math. Z.* **124** (1972), 261-265.

[LiSa]   M. Liebeck and J. Saxl, Minimal degrees of primitive permutation groups, with an application to monodromy groups of covers of Riemann surfaces, *Proc. London Math. Soc.* (3) **63** (1991), no. 2, 266-314.

[LiSh]   M. Liebeck and A. Shalev, Simple groups, permutation groups, and probability, *J. Amer. Math. Soc.* **12** (1999), no. 2, 497-520.

[LiWP]   M. Liebeck and C. Wayman Purvis, On the genus of a finite classical group, *Bull. London Math. Soc.* **29** (1997), no. 2, 159-164.

[MaVo]   K. Magaard and H. Völklein, The monodromy group of a function on a general curve, *Israel J. Math.* **141** (2004), 355–368.

[Mu]   P. Müller, personal communication, 1998.

[Ne1]   M. Neubauer, On monodromy groups of fixed genus, *J. Algebra* **153** (1992), no. 1, 215-261.

[Ne2]   M. Neubauer, On primitive monodromy groups of genus zero and one, I, *Comm. Algebra* **21** (1993), no. 3, 711-746.

[Re]   R. Ree, A theorem on permutations, *J. Combinatorial Theory Ser. A* **10** (1971), 174-175.

[Sa]   B. Sagan, *The Symmetric Group*, Wadsworth, Belmont, CA, 1991.

[Sc]   L. L. Scott, Matrices and cohomology, *Ann. of Math.* (2) **105** (1977), no. 3, 473-492.

[Sh1]   T. Shih, A note on groups of genus zero, *Comm. Algebra* **19** (1991), no. 10, 2813-2826.

[Sh2]   T. Shih, *Bounds of fixed point ratios of permutation representations of $GL_n(q)$ and groups of genus zero*, Ph.D. Thesis, California Institute of Technology, 1990.

[Vo]   H. Völklein, *Groups as Galois Groups*, Cambridge University Press, Cambridge, 1996.

Editorial Information

To be published in the *Memoirs*, a paper must be correct, new, nontrivial, and significant. Further, it must be well written and of interest to a substantial number of mathematicians. Piecemeal results, such as an inconclusive step toward an unproved major theorem or a minor variation on a known result, are in general not acceptable for publication.

Papers appearing in *Memoirs* are generally at least 80 and not more than 200 published pages in length. Papers less than 80 or more than 200 published pages require the approval of the Managing Editor of the Transactions/Memoirs Editorial Board.

As of May 31, 2007, the backlog for this journal was approximately 15 volumes. This estimate is the result of dividing the number of manuscripts for this journal in the Providence office that have not yet gone to the printer on the above date by the average number of monographs per volume over the previous twelve months, reduced by the number of volumes published in four months (the time necessary for preparing a volume for the printer). (There are 6 volumes per year, each usually containing at least 4 numbers.)

A Consent to Publish and Copyright Agreement is required before a paper will be published in the *Memoirs*. After a paper is accepted for publication, the Providence office will send a Consent to Publish and Copyright Agreement to all authors of the paper. By submitting a paper to the *Memoirs*, authors certify that the results have not been submitted to nor are they under consideration for publication by another journal, conference proceedings, or similar publication.

Information for Authors

Memoirs are printed from camera copy fully prepared by the author. This means that the finished book will look exactly like the copy submitted.

Initial submission. The AMS uses Centralized Manuscript Processing for initial submissions. Authors should submit a PDF file using the Initial Manuscript Submission form found at www.ams.org/cgi-bin/peertrack/submission.pl, or send one copy of the manuscript to the following address: Centralized Manuscript Processing, MEMOIRS OF THE AMS, 201 Charles Street, Providence, RI 02904-2294 USA. If a paper copy is being forwarded to the AMS, indicate that it is for it Memoirs and include the name of the corresponding author, contact information such as email address or mailing address, and the name of an appropriate Editor to review the paper (see the list of Editors below).

The paper must contain a *descriptive title* and an *abstract* that summarizes the article in language suitable for workers in the general field (algebra, analysis, etc.). The *descriptive title* should be short, but informative; useless or vague phrases such as "some remarks about" or "concerning" should be avoided. The *abstract* should be at least one complete sentence, and at most 300 words. Included with the footnotes to the paper should be the 2000 *Mathematics Subject Classification* representing the primary and secondary subjects of the article. The classifications are accessible from www.ams.org/msc/. The list of classifications is also available in print starting with the 1999 annual index of *Mathematical Reviews*. The Mathematics Subject Classification footnote may be followed by a list of *key words and phrases* describing the subject matter of the article and taken from it. Journal abbreviations used in bibliographies are listed in the latest *Mathematical Reviews* annual index. The series abbreviations are also accessible from www.ams.org/publications/. To help in preparing and verifying references, the AMS offers MR Lookup, a Reference Tool for Linking, at www.ams.org/mrlookup/.

Electronically prepared manuscripts. The AMS encourages electronically prepared manuscripts, with a strong preference for $\mathcal{AMS}$-LATEX. To this end, the Society has prepared $\mathcal{AMS}$-LATEX author packages for each AMS publication. Author packages include instructions for preparing electronic manuscripts, samples, and a style file that generates the particular design specifications of that publication series. Though $\mathcal{AMS}$-LaTeX is the highly preferred format of TeX, author packages are also available in $\mathcal{AMS}$-TeX.

Authors may retrieve an author package from the AMS website starting from www.ams.org/tex/ or via FTP to ftp.ams.org (login as anonymous, enter username as password, and type cd pub/author-info). The *AMS Author Handbook* and the *Instruction Manual* are available in PDF format following the author packages link from www.ams.org/tex/. The author package can also be obtained free of charge by sending email to tech-support@ams.org (Internet) or from the Publication Division, American Mathematical Society, 201 Charles St., Providence, RI 02904-2294, USA. When requesting an author package, please specify $\mathcal{AMS}$-LaTeX or $\mathcal{AMS}$-TeX and the publication in which your paper will appear. Please be sure to include your complete mailing address.

After acceptance. The final version of the electronic file should be sent to the Providence office (this includes any TeX source file, any graphics files, and the DVI or PostScript file) immediately after the paper has been accepted for publication.

Before sending the source file, be sure you have proofread your paper carefully. The files you send must be the EXACT files used to generate the proof copy that was accepted for publication. For all publications, authors are required to send a printed copy of their paper, which exactly matches the copy approved for publication, along with any graphics that will appear in the paper.

Accepted electronically prepared files can be submitted via the web at www.ams.org/submit-book-journal/, sent via FTP, or sent on CD-Rom or diskette to the Electronic Prepress Department, American Mathematical Society, 201 Charles Street, Providence, RI 02904-2294 USA. TeX source files, DVI files, and PostScript files can be transferred over the Internet by FTP to the Internet node ftp.ams.org (130.44.1.100). When sending a manuscript electronically via CD-Rom or diskette, please be sure to include a message identifying the paper as a Memoir.

Electronically prepared manuscripts can also be sent via email to pub-submit@ams.org (Internet). In order to send files via email, they must be encoded properly. (DVI files are binary and PostScript files tend to be very large.)

Electronic graphics. Comprehensive instructions on preparing graphics are available at www.ams.org/jourhtml/. A few of the major requirements are given here.

Submit files for graphics as EPS (Encapsulated PostScript) files. This includes graphics originated via a graphics application as well as scanned photographs or other computer-generated images. If this is not possible, TIFF files are acceptable as long as they can be opened in Adobe Photoshop or Illustrator. No matter what method was used to produce the graphic, it is necessary to provide a paper copy to the AMS.

Authors using graphics packages for the creation of electronic art should also avoid the use of any lines thinner than 0.5 points in width. Many graphics packages allow the user to specify a "hairline" for a very thin line. Hairlines often look acceptable when proofed on a typical laser printer. However, when produced on a high-resolution laser imagesetter, hairlines become nearly invisible and will be lost entirely in the final printing process.

Screens should be set to values between 15% and 85%. Screens which fall outside of this range are too light or too dark to print correctly. Variations of screens within a graphic should be no less than 10%.

Inquiries. Any inquiries concerning a paper that has been accepted for publication should be sent to memo-query@ams.org or directly to the Electronic Prepress Department, American Mathematical Society, 201 Charles St., Providence, RI 02904-2294 USA.

Editors

This journal is designed particularly for long research papers, normally at least 80 pages in length, and groups of cognate papers in pure and applied mathematics. Papers intended for publication in the *Memoirs* should be addressed to one of the following editors. The AMS uses Centralized Manuscript Processing for initial submissions to AMS journals. Authors should follow instructions listed on the Initial Submission page found at www.ams.org/memo/memosubmit.html.

Algebra to ALEXANDER KLESHCHEV, Department of Mathematics, University of Oregon, Eugene, OR 97403-1222; email: ams@noether.uoregon.edu

Algebraic geometry and its application to MINA TEICHER, Emmy Noether Research Institute for Mathematics, Bar-Ilan University, Ramat-Gan 52900, Israel; email: teicher@macs.biu.ac.il

Algebraic geometry to DAN ABRAMOVICH, Department of Mathematics, Brown University, Box 1917, Providence, RI 02912; email: amsedit@math.brown.edu

Algebraic number theory to V. KUMAR MURTY, Department of Mathematics, University of Toronto, 100 St. George Street, Toronto, ON M5S 1A1, Canada; email: murty@math.toronto.edu

Algebraic topology to ALEJANDRO ADEM, Department of Mathematics, University of British Columbia, Room 121, 1984 Mathematics Road, Vancouver, British Columbia, Canada V6T 1Z2; email: adem@math.ubc.ca

Combinatorics to JOHN R. STEMBRIDGE, Department of Mathematics, University of Michigan, Ann Arbor, Michigan 48109-1109; email: FRS@umich.edu

Complex analysis and harmonic analysis to ALEXANDER NAGEL, Department of Mathematics, University of Wisconsin, 480 Lincoln Drive, Madison, WI 53706-1313; email: nagel@math.wisc.edu

Differential geometry and global analysis to LISA C. JEFFREY, Department of Mathematics, University of Toronto, 100 St. George St., Toronto, ON Canada M5S 3G3; email: jeffrey@math.toronto.edu

Dynamical systems and ergodic theory to AMIE WILKINSON, Department of Mathematics, Northwestern University, 2033 Sheridan Road, Evanston, IL 60208-2730; email: transactions@math.northwestern.edu

Functional analysis and operator algebras to DIMITRI SHLYAKHTENKO, Department of Mathematics, University of California, Los Angeles, CA 90095; email: shlyakht@math.ucla.edu

Geometric analysis to WILLIAM P. MINICOZZI II, Department of Mathematics, Johns Hopkins University, 3400 N. Charles St., Baltimore, MD 21218; email: trans@math.jhu.edu

Geometric analysis to MLADEN BESTVINA, Department of Mathematics, University of Utah, 155 South 1400 East, JWB 233, Salt Lake City, Utah 84112-0090; email: bestvina@math.utah.edu

Harmonic analysis, representation theory, and Lie theory to ROBERT J. STANTON, Department of Mathematics, The Ohio State University, 231 West 18th Avenue, Columbus, OH 43210-1174; email: stanton@math.ohio-state.edu

Logic to STEFFEN LEMPP, Department of Mathematics, University of Wisconsin, 480 Lincoln Drive, Madison, Wisconsin 53706-1388; email: lempp@math.wisc.edu

Partial differential equations to GUSTAVO PONCE, Department of Mathematics, South Hall, Room 6607, University of California, Santa Barbara, CA 93106; email: ponce@math.ucsb.edu

Partial differential equations and dynamical systems to PETER POLACIK, School of Mathematics, University of Minnesota, Minneapolis, MN 55455; email: polacik@math.umn.edu

Probability and statistics to KRZYSZTOF BURDZY, Department of Mathematics, University of Washington, Box 354350, Seattle, Washington 98195-4350; email: burdzy@math.washington.edu

Real analysis and partial differential equations to DANIEL TATARU, Department of Mathematics, University of California, Berkeley, Berkeley, CA 94720; email: tataru@math.berkeley.edu

All other communications to the editors should be addressed to the Managing Editor, ROBERT GURALNICK, Department of Mathematics, University of Southern California, Los Angeles, CA 90089-1113; email: guralnic@math.usc.edu.

Titles in This Series

887 **Charlotte Wahl,** Noncommutative Maslov index and eta-forms, 2007

886 **Robert M. Guralnick and John Shareshian,** Symmetric and alternating groups as monodromy groups of Riemann surfaces I: Generic covers and covers with many branch points, 2007

885 **Jae Choon Cha,** The structure of the rational concordance group of knots, 2007

884 **Dan Haran, Moshe Jarden, and Florian Pop,** Projective group structures as absolute Galois structures with block approximation, 2007

883 **Apostolos Beligiannis and Idun Reiten,** Homological and homotopical aspects of torsion theories, 2007

882 **Lars Inge Hedberg and Yuri Netrusov,** An axiomatic approach to function spaces, spec tral synthesis and Luzin approximation, 2007

881 **Tao Mei,** Operator valued Hardy spaces, 2007

880 **Bruce C. Berndt, Geumlan Choi, Youn-Seo Choi, Heekyoung Hahn, Boon Pin Yeap, Ae Ja Yee, Hamza Yesilyurt, and Jinhee Yi,** Ramanujan's forty identities for Rogers-Ramanujan functions, 2007

879 **O. García-Prada, P. B. Gothen, and V. Muñoz,** Betti numbers of the moduli space of rank 3 parabolic Higgs bundles, 2007

878 **Alessandra Celletti and Luigi Chierchia,** KAM stability and celestial mechanics, 2007

877 **María J. Carro, José A. Raposo, and Javier Soria,** Recent developments in the theory of Lorentz spaces and weighted inequalities, 2007

876 **Gabriel Debs and Jean Saint Raymond,** Borel liftings of Borel sets: Some decidable and undecidable statements, 2007

875 **C. Krattenthaler and T. Rivoal,** Hypergéométrie et fonction zêta de Riemann, 2007

874 **Sonia Natale,** Semisolvability of semisimple Hopf algebras of low dimension, 2007

873 **A. J. Duncan,** Exponential genus problems in one-relator products of groups, 2007

872 **Anthony V. Geramita, Tadahito Harima, Juan C. Migliore, and Yong Su Shin,** The Hilbert function of a level algebra, 2007

871 **Pascal Auscher,** On necessary and sufficient conditions for L^p-estimates of Riesz transforms associated to elliptic operators on $\mathbb{R}^n$ and related estimates, 2007

870 **Takuro Mochizuki,** Asymptotic behaviour of tame harmonic bundles and an application to pure twistor D-modules, Part 2, 2007

869 **Takuro Mochizuki,** Asymptotic behaviour of tame harmonic bundles and an application to pure twistor D-modules, Part 1, 2007

868 **Gelu Popescu,** Entropy and multivariable interpolation, 2006

867 **Vilmos Totik,** Metric properties of harmonic measures, 2006

866 **William Craig,** Semigroups underlying first-order logic, 2006

865 **Nathanial P. Brown,** Invariant means and finite representation theory of $C*$-algebras, 2006

864 **John M. Lee,** Fredholm operators and Einstein metrics on conformally compact manifolds, 2006

863 **M. Lübke and A. Teleman,** The Universal Kobayashi-Hitchin correspondence on Hermitian manifolds, 2006

862 **Alberto Canonaco,** The Beilinson complex and canonical rings of irregular surfaces, 2006

861 **Leon A. Takhtajan and Lee-Peng Teo,** Weil-Petersson metric on the universal Teichmüller space, 2006

860 **Thomas M. Fiore,** Pseudo limits, biadjoints and pseudo algebras: Categorical foundations of conformal field theory, 2006

859 **N. Arcozzi, R. Rochberg, and E. Sawyer,** Carleson measures and interpolating sequences for Besov spaces on complex balls, 2006

TITLES IN THIS SERIES

858 **Enrico Valdinoci, Berardino Sciunzi, and Vasile Ovidiu Savin,** Flat level set regularity of p-Laplace phase transitions, 2006

857 **Donatella Danielli, Nocola Garofalo, and Duy-Minh Nhieu,** Non-doubling Ahlfors measures, perimeter measures, and the characterization of the trace spaces of Sobolev functions in Carnot-Carathéodory spaces, 2006

856 **Vladimir Bolotnikov and Harry Dym,** On boundary interpolation for matrix valued Schur functions, 2006

855 **Yevgenia Kashina, Yorck Sommerhäuser, and Yongchang Zhu,** On higher Frobenius-Schur indicators, 2006

854 **Noam Greenberg,** The role of true finiteness in the admissible recursively enumerable degrees, 2006

853 **Joachim Krieger,** Stability of spherically symmetric wave maps, 2006

852 **Viorel Barbu, Irena Lasiecka, and Roberto Triggiani,** Tangential boundary stabilization of Navier-Stokes equations, 2006

851 **Jie Wu,** On maps from loop suspensions to loop spaces and the shuffle relations on the Cohen groups, 2006

850 **Siegfried Echterhoff, S. Kaliszewski, John Quigg, and Iain Raeburn,** A categorical approach to imprimitivity theorems for C^*-dynamical systems, 2006

849 **Katsuhiko Kuribayashi, Mamoru Mimura, and Tetsu Nishimoto,** Twisted tensor products related to the cohomology of the classifying spaces of loop groups, 2006

848 **Bob Oliver,** Equivalences of classifying spaces completed at the prime two, 2006

847 **Eric T. Sawyer and Richard L. Wheeden,** Hölder continuity of weak solutions to subelliptic equations with rough coefficients, 2006

846 **Victor Beresnevich, Detta Dickinson, and Sanju Velani,** Measure theoretic laws for lim–sup sets, 2006

845 **Ehud Friedgut, Vojtech Rödl, Andrzej Ruciński, and Prasad V. Tetali,** A Sharp threshold for random graphs with a monochromatic triangle in every edge coloring, 2006

844 **Amadeu Delshams, Rafael de la Llave, and Tere M. Seara,** A geometric mechanism for diffusion in Hamiltonian systems overcoming the large gap problem: Heuristics and rigorous verification on a model, 2006

843 **Denis V. Osin,** Relatively hyperbolic groups: Intrinsic geometry, algebraic properties, and algorithmic problems, 2006

842 **David P. Blecher and Vrej Zarikian,** The calculus of one-sided M-ideals and multipliers in operator spaces, 2006

841 **Enrique Artal Bartolo, Pierrette Cassou-Noguès, Ignacio Luengo, and Alejandro Melle Hernández,** Quasi-ordinary power series and their zeta functions, 2005

840 **Sławomir Kołodziej,** The complex Monge-Ampère equation and pluripotential theory, 2005

839 **Mihai Ciucu,** A random tiling model for two dimensional electrostatics, 2005

838 **V. Jurdjevic,** Integrable Hamiltonian systems on complex Lie groups, 2005

837 **Joseph A. Ball and Victor Vinnikov,** Lax-Phillips scattering and conservative linear systems: A Cuntz-algebra multidimensional setting, 2005

836 **H. G. Dales and A. T.-M. Lau,** The second duals of Beurling algbras, 2005

835 **Kiyoshi Igusa,** Higher complex torsion and the framing principle, 2005

834 **Kenîchi Ohshika,** Kleinian groups which are limits of geometrically finite groups, 2005

For a complete list of titles in this series, visit the AMS Bookstore at **www.ams.org/bookstore/**.

WITHDRAWN